I

II

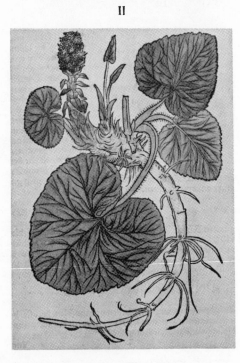

III

IV

HANDBOOK OF PLANT AND FLORAL ORNAMENT

SELECTED FROM THE HERBALS OF THE SIXTEENTH CENTURY, AND EXHIBITING THE FINEST EXAMPLES OF PLANT-DRAWING FOUND IN THOSE RARE WORKS, WHETHER EXECUTED IN WOOD-CUTS OR IN COPPER-PLATE ENGRAVINGS, ARRANGED FOR THE USE OF THE DECORATOR WITH SUPPLEMENTARY ILLUSTRATIONS AND SOME REMARKS ON THE USE OF PLANT-FORM IN DESIGN BY **RICHARD G. HATTON**

FORMERLY TITLED: *The Craftsman's Plant-Book: or Figures of Plants*

DOVER PUBLICATIONS, INC.
NEW YORK

Published in Canada by General Publishing Company, Ltd., 30 Lesmill Road, Don Mills, Toronto, Ontario.
Published in the United Kingdom by Constable and Company, Ltd., 10 Orange Street, London WC 2.

This Dover edition, first published in 1960, is an unabridged and unaltered republication of the work originally published in 1909 under the title *The Craftsman's Plant-Book.*
The frontispiece which originally appeared in color is now reproduced in black and white.
This work is republished by special arrangement with Chapman and Hall, Limited.

Standard Book Number: 486-20649-1
Library of Congress Catalog Card Number: 60-50719

Manufactured in the United States of America
Dover Publications, Inc.
180 Varick Street
New York, N. Y. 10014

PREFACE

THE object of this publication is to render available to designers and plant-lovers the best of the engraved drawings of plants which have made the Herbals of the sixteenth century famous.

Executed while botanical science was still in its infancy, these old drawings are not likely to satisfy those persons who look always for the marks by which genus and species are distinguished.

But to the artist, and particularly to the designer or craftsman, these figures can hardly fail to appeal. Full of decorative suggestion, and fine examples of treatment, they present the plants as they are known to ordinary people, with their character usually admirably expressed.

There is a revival at the present time of the use of plant form in design; not, indeed, of the use of later Victorian days, in which Nature was regarded merely as a store-house of form, but of that of the days of Elizabeth, James I and Charles I, when men used flowers and plants in decoration, because they knew and loved the plants themselves, and delighted to represent them. Of this present-day revival William Morris was the pioneer; and he looked back to those earlier times.

In such decorative work the figures of plants in the old herbals are of the greatest value. In many cases they are designs ready-done, and can be appropriated wholesale. Observe, for instance, the use made by Sir E. Burne-Jones in the design for the cover of Ruskin's *Studies in Both Arts* of figures of the Purple Wind-flower and White Lily, reproduced in this book.

But the old herbals are unwieldy and costly; and the aim of the compiler has therefore been to collect from all of them the best and most helpful figures into a handy compass.

The best use of these drawings will doubtless be made by those who also study the living plants themselves; and it cannot be too strongly impressed upon the student, that these ancient figures owe their merit very largely to their being drawings, at first hand, from nature. They were called into existence by a demand for true drawing—for "living figures," as Brunfels called them, and it is because, with all their peculiarities of arrangement—their square heads, and so forth—they are living, that they have such charm, and are of so much value to the artist.

When they were made, the external appearance of the plant was the main guide to its identity, as it must be in all artistic work, and many details, even of external form, were consequently overlooked or erroneously rendered. These deficiencies we now-a-days feel we must supply. And therefore it has seemed incumbent upon the compiler to add a number of drawings of his own. In all cases where a reference is not given the drawings are his. Some few figures are from Lindley's *Vegetable Kingdom.*

Further, he has felt that the figures would be most useful if arranged according to the Natural Orders, and he has prefixed to each Order a brief description, such as he thinks will be helpful to the designer in representing the plants. These descriptions he does not pretend have any claim to render the work a botanical one, any more than he pretends to be a botanist. But they may serve their purpose, and may, furthermore, induce the reader to take up the study of botany.

It is with much regret that the compiler finds himself obliged, as he prepares the sheets for the press, to cast aside the descriptions of the individual plants which he had prepared, and which he now reduces to the bare statement of colour, size and season. But the space occupied even by short descriptions is better given, in a work of this character, to illustrations. As to the range of plants presented, it has to be said that it has been impossible even to mention all the native British plants, still less all the garden plants. At the same time, many plants are represented by more than one figure—for, to the artist, a different treatment is as good as a different plant. Of the figures of Flax, for instance, which of them could have been omitted? Sometimes the ancient figures, however true in a general sense, were not worth reproducing. Nevertheless, the figures given fairly represent the whole range of plant-life.

The books from which the figures have been reproduced commence with Brunfels (1530) and end with Crispian van de Passe (1614). The later works which have been used—Johnson's Gerarde (1633) and Zwinger (1744) were illustrated with blocks cut long before—except for a few instances.

PREFACE

The range 1530–1614 covers all the fine books. The earlier, and still more costly, *Grete Herbal* and *Ortus Sanitatis* have quaint and curious cuts, which, though interesting, are readily eclipsed for our purposes by the later works. And the books which are later than those here used lack charm and truth. Even Parkinson's *Paradisus* (1629) must be regarded as a crude performance. Many of its figures are readily recognized as being out of Crispian van de Passe, and from the same sources as those in Johnson.

A list of the books from which figures have been taken is given on a later page. They form a most interesting record of the rise of the study of plants. They begin as Herbals, or Books of Health, then become General Histories of Plants, and finally become Gardens.

They begin, that is, as books of medicine with figures to secure identification; then the order becomes reversed, and the plants are first sought and their "virtues" then compiled; and lastly, cultivation is adopted, and varieties for the pleasure garden are dealt with.

Nevertheless, the period of the fine drawings closed before botanical science arose in anything like its modern form. And when it arose the day had come rather for dissections than for " figures," and there were no longer produced the elaborate representations which were thought necessary in the time of Henry VIII and Elizabeth.

Of the high merit of these figures it is hardly necessary to say more here than that it was the result of a fortunate combination of circumstances which cannot perhaps recur.

It was the custom in the old days to colour the figures. This was sometimes done by the publishers. A few examples from a coloured copy of Bock are given in the frontispiece. This present book is printed on paper which will take colour well; and it is a good plan for a reader to colour a figure when he finds the living plant.

Some assistance in colouring the figures will be found by consulting the hand-coloured books issued between, say, 1780 and 1840. These books are referred to in the Appendix. What one has particularly to remember is that the colouring should be graduated, and not put on flatly, and that different hues should be blended together; and also, that it is not unusual for the colour of the flowers to tinge the stalks, and perhaps the leaves, near them.

Thanks are particularly due to Colonel H. B. Hanna for permission to reproduce a fine Indian drawing in which there is both plant- and animal-drawing of an extremely high order. Colonel Hanna says that the drawing was probably made in the time of the Emperor Akbar, about 1570, when art in India was at its zenith. This date is practically the same as that of the best European drawings of plants. But it may be doubted whether such beautiful drawings of animals were done so early in Europe.

When, one ventures to ask, will people recognize that India had an art which rises superior to anything produced by other Eastern nations, and is unsurpassed, in its kind, by anything that Europe has to show?

The tree in the foreground of the drawing, whose leaves a deer is nibbling, is probably an acacia: the flowers on the right side are composites—sunflowers; and the tree in the left corner near the bounding deer and hunting leopard is apparently the dākh, a tree whose blossom is a bright crimson and which when in flower droops its leaves. A forest of these trees in full blossom is a most dazzling and beautiful sight.

The Iris will be seen to be very ably rendered, the beard being shown. The upper part of the picture has unfortunately been damaged, and the mass of foliage which once crowned it is now lost.

CONTENTS

ix

HANDBOOK OF PLANT
AND FLORAL ORNAMENT

CHAPTER I

THE OLD HERBALS

OF all the old Herbals, Gerarde is the best known. The title of the book is *The Herball, or Generall Historie of Plantes*, gathered by John Gerarde of London, Master in Chirurgerie. It was printed at London in 1597. This herbal was "very much enlarged, and amended" by Thomas Johnson, Citizen and Apothecary, of London. This second edition was published in 1633, and a second impression was issued in 1636. It is the version of Johnson which is the Gerarde commonly known, as it is a decidedly less scarce book than the first edition, and it is certainly much enlarged and amended.

In a learned preface Johnson reviews the various writers on plants, coming finally to John Gerarde, whose work he was revising, and whose popularity was apparently so great at the time that it was better worth Johnson's while to re-issue and revise his book than to produce one of his own. He pours some contempt upon Gerarde, whom he accuses of having built his book upon a manuscript which had been left unpublished by a Dr. Priest, whose labours Gerarde made no more acknowledgment of than to say he had heard of them.

Whatever were or were not the abilities of Gerarde, it is commonly allowed that he gave a great spur to the culture of plants in England, and to the observation of them. His name is now certainly fading somewhat from the pages of botanical science—nay, it has gone altogether, but for an occasional reference to him as one of the pioneers.

It is, however, almost as high a distinction with posterity to be among the rarities in the world of books as to be mentioned in the pages of science —and Gerarde fetches a good price. The book owes its popularity, no doubt, as much to its quaintness as to its woodcut illustrations. One truly does not get as fine a feast of strange cures, and magical medicines, as might be expected, but in a quiet way there is plenty to amuse one.

Johnson generally prints Gerarde's text precisely as it stands, adding his corrections. He speaks often of the wrong figures being placed here and there, and many of the mistakes are obvious enough. But then the sets of illustrations used by Gerarde and by Johnson are not the same. Those used by Gerarde had been prepared for Theodorus Tabernæmontanus, who published his *Neuw Kreuterbuch* at Frankfort in 1588. These blocks were brought over for the printing of Gerarde's Herbal, and were then taken back again and used in subsequent issues of Tabernæmontanus.

Johnson twits Gerarde upon the difficulties he had in fitting the figures of Tabernæmontanus to his own work, which he claimed was merely a " crib " from Dodonæus and Lobel. Nevertheless, Johnson himself escapes the difficulty, for, as he tells us, for his edition there were used the blocks of Dodonæus, Lobel, and Clusius. The works of these three botanists were mostly printed at Antwerp, where Plantin, the printer, got together a large collection of blocks, and became the recognized printer of plant-books, just as Egenolph of Frankfort had been, thirty or forty years earlier. One is not surprised, therefore, to find the blocks much alike in style and execution. Some owed their superior accuracy to the careful draughtsmanship of Clusius himself, but all had the same qualities of boldness and simplicity, marred by a little clumsiness, and rather devoid of beauty and gracefulness. The blocks, having been used in one or another work, were used, apparently for the last time, in Johnson's edition of Gerarde.

The original edition of Gerarde is a thick folio of over fourteen hundred pages, twelve and a half inches high and eight and a half wide. It has a copperplate title-page and a portrait of Gerarde, both by William Rogers. It has about 2,190 cuts, of which sixteen were new, and some, perhaps a hundred, were from Lobel. The figures are generally four and three-quarters inches high and three wide.

In the figures given in this selection all those marked G are from the original edition of Gerarde, and are cuts from Tabernæmontanus. Those marked J are from Johnson's version, and represent the work of the three writers named together above.

They were very skilful in those days at reproducing woodcuts. They either traced them down, or had some means of laying an impression on the block, and recutting it. This was sometimes so skilfully done that it is only by carefully comparing every part of the impressions that one can find by some alteration of the lines that the impressions are not from the same block.

Credit must be given sometimes for very excellent redrawing. Notably in the *Historia Generalis Plantarum* of Dalechamps, which was published at Lyons in 1586, there are some excellent examples. In them the rectangular shape of the block is usually taller and narrower than are those of the impressions from which they are copied. Some of the work in this book is extremely well done, and the cutting is of a very high order. This is a pleasant fact to the book-lover, because the books turned out at Lyons had often very poorly and casually executed woodcuts.

Some of the cuts in Tabernæmontanus are excellently engraved. For fineness of execution they sometimes reached a very high level of attainment, and perhaps were not excelled by any others. But they are chiefly attractive for their obvious symmetrical arrangement. This of course makes them very "decorative." It is possible sometimes to trace them to their origin, in one of the other books, and one finds that the draughtsman for Tabernæmontanus has not scrupled to make a stem branch into two, if by so doing it would fill his block. Very commonly he stiffened the plant out, and gave all its branches the same height, and so on, rendering the figure very suitable for use as ornament, but losing its natural appearance. He thus made his drawings suggestive to the designer, and one cannot but be glad that he adopted this style, for it certainly shows the craftsman how to proceed.

In 1530 Otto Brunfels issued at Strasburg his *Herbarum Vivae Eicones*, or Living Figures of Plants. The title of the work is important. It indicates that the figures of plants which were then in use were not drawn from nature, and it indicates also, what is more important, that men were beginning to carefully observe the plants with their own eyes.

The illustrations in Brunfels are unequal in merit, or perhaps they are unequal in the engraving. Many of them reach at once a very high level; and it may be doubted whether the beautiful figures of Fuchsius excel them. They are not conventional at all in arrangement. There is, as usual, much beauty of disposition, but there is no attempt to render the drawings balanced or symmetrical. They were done by Hans Weiditz, an engraver of Strasburg, some of whose work has been ascribed to Burgkmair and to Dürer. A considerable account of him is given in Bryan's *Dictionary of Painters, etc.*

The books which Brunfels superseded are quaint, rare, costly volumes with a profusion of coarsely executed cuts. The chief is the *Ortus Sanitatis*, the first edition of which was issued at Mainz in 1475. Another is the *Buch der Natur*, printed at Augsburg in 1482, and another is the *Grete Herball*, published in London in 1526. The figures in these books have, as a rule, little to recommend them. One might expect that the simple small drawings would be very suggestive for such ornaments as badges, and similar little details. But they are not. Though one can often see what they are meant to represent, they lack real correspondence with natural forms, and do not compensate for this by any beauty of design. It is not unlikely that these simple cuts are copies of the drawings which illustrated some of the manuscripts of the Middle Ages. There were certain ancient writers on plants—Theophrastus, Dioscorides, and Pliny are the greatest names. Their works

I.—VERVAIN.
From *De Viribus Herbarum*—MACER.
C. 1490.

II.—MUGWORT.
From *De Viribus Herbarum*—MACER.
C. 1490.

were copied in manuscript, often, if not generally, with drawings of the plants beside the descriptions. The drawings are in very many cases rather in the nature of embellishments of the page than really serviceable adjuncts to the text.

These early drawings are quite ideographic; not in any sense imitative. Usually they present four or five leaves of the form supposed to be characteristic of the plant, arranged in symmetrical fashion upon an upright central stem. Generally the leaves are opposite, not because they are opposite in nature but because it was, and still is, the habit of man's mind to prefer a symmetrical arrangement if he sees no reason for departing from it.

In an edition of Macer—*De Viribus Herbarum*—in the British Museum, there are cuts which might well be repetitions of drawings in manuscripts. Whether any manuscripts of books about plants exist in which plant drawings occur of the excellent character sometimes seen in the later manuscripts, I do not know—certainly the early woodcuts bear no resemblance to such careful still-life painting as the borders of many a Book of Hours exhibit.

Brunfels, then, it was who started a new era in the illustration of books about plants. We are hardly concerned with the subject-matter of the books which poured from the presses during the sixteenth century without ceasing. They were at first republications of the wisdom of the ancient writers, and were in a measure compilations of the " virtues " of plants in medicine, collected out of the Greek, Latin and Arabian authors. There were generally given short descriptions of the plants, and figures to assist the reader in getting hold of the right plants. Very much difficulty was naturally found in determining which were the plants referred to by the ancient authorities. Divergencies between the floras of the different countries in which the works had originally been composed and were now being studied added greatly to the difficulties which these early botanists encountered. Gradually they felt obliged to examine and describe all the plants they found, and the herbals, or physic-books, developed into general histories of plants. In these books one sees botanical science gradually arising. Observation of one set of facts after another is revealed as book follows book, but the era of the fine herbals closed before the pistils and stamens were properly examined.

The illustrations in the herbals fall into schools. The first school is that of Brunfels, and perhaps it is limited to the work of Brunfels himself. The cuts produced under Egenolph at Frankfort might perhaps be classed with them, but the crisp, angular, and yet sure and obedient line of Brunfels is hardly followed in the flowing, easy, almost brush-like and indeed rather crude work of Egenolph, who published many works with important wood-cut illustrations. The next school is that of Fuchsius and Bock. Fuchsius published his fine *Historia Stirpium* at Basle in 1542. The figures are larger even than those of Brunfels, and are about twelve or thirteen inches high. These magnificent drawings are in outline only, for Fuchsius would not permit any shading, lest the form should be confused, and the lines are cut to an unusual narrowness.

These figures of Fuchsius immediately set to work a whole bevy of pirates, of whose thievish hands Fuchsius complained in the preface to the small versions of the figures which he issued in 1545. These small cuts are indeed very beautifully executed, and some of them are better than the large drawings from which they were taken. This improvement is due to the parts having to be crowded more closely together, so that the reduction might not be greater than was actually or absolutely necessary. Egenolph was one of the thieves who stole Fuchsius's figures, and the simple, brush-like drawings to which we have alluded are often simplifications of the cuts in Fuchsius. The whole lot was bagged and re-engraved by Birckmann of Cologne, and were used by him in printing the *New Herbal* of William Turner (A.D. 1551), a work of which Englishmen may justly be proud. Turner was an honest observer, and had a freer mind as regards the properties of the subjects he examined than many of the writers of his time. Any book on plants published by Birckmann is almost sure to have in it these copies of the small cuts of Fuchsius. Some of them are fairly well rendered, but many are merely chopped out of the wood. The same set of blocks were used in Lyte's edition of Dodonæus.

Hieronymus Tragus, or Bock, issued his *Kreutter Buch* at Strasburg in 1546. The figures were drawn by David Kandel. Many of them are copied from Fuchsius and Brunfels, but many are original. They exhibit the plant in its natural growth, without any attempt at conventional arrangement. It is thought that Kandel cut them on the wood himself, and they are certainly admirable pieces of engraving.

Many of these old herbals are coloured by hand. The colouring was often contemporary. It generally lessens the price of the book on the book market ; but I cannot but think that it was intended that the copies should be coloured and issued in that form by the publisher. Often the colour is put on carelessly, and the form is obscured. Nevertheless, the effect is enhanced, and one can hardly believe that Brunfels would have called his book "living figures of plants," if they were not intended to be coloured. The figures in Fuchsius look as if they want colouring. Some contend that one of the three artists whose portraits are given at the back was the painter of the printed sheets. It seems to me that Albertus Meyer is drawing the plant on paper, that Henricus Füllmaurer is drawing upon the block from a drawing, perhaps one of Meyer's. The third artist is Vitus Rudolphus Speckle, the engraver. One cannot but note that the engraver, or draughtsman, of the large cuts of Matthiolus had the same initial to his surname—W. S.

A copy of Brunfels in the British Museum is admirably coloured, much act being added with the brush, but the coloured copies of Fuchsius in the Museum are poorly done.

Another style of drawing is that seen in the Commentaries upon the Six Books of Dioscorides by Matthiolus, which was issued at Venice in 1554. It was re-issued with more figures in 1558. They exhibit more arrangement than do the cuts in Fuchsius, and are a little deficient in variety of touch. One of their charms is in the shading, which shows a pleasant mastery of parallel lines. These drawings must be classed with the magnificent large blocks which ennobled the edition of Matthiolus which was issued in Venice in 1565. I have not seen any opinion as to who was the draughtsman or the engraver of these cuts, but by the initials upon the drawing of the Orange—W. S.—it would seem that either one or the other was a German.

Another school of engraving was that at Lyons, where many books with woodcuts were turned out. The style is as a rule rather ragged and casual, as if the draughtsmen did not worry about their corners so much as did the Germans. But in the large *Historia Generalis Plantarum* of Dalechamps, published at Lyons in 1586, which has already been referred to, there is as good cutting as in any of the herbals. In this book are included a very large selection of figures from Fuchsius, Matthiolus, Dodonæus, Lobel, Clusius, and also from other botanists who dealt with plants from distant countries. Some of these are quaint and curious, and are chiefly interesting on that score.

Mention has already been made of Dodonæus, Lobel, and Clusius, at the beginning of this chapter. They form a decided school of plant drawing ; and Tabernæmontanus has also been noticed.

Next to claim attention is the important series of blocks which were prepared for Gesner, but which he did not live to use, and which were employed by Camerarius in his *Epitome Matthioli* published at Frankfort in 1586. This book did not contain all Gesner's cuts, and a larger number is to be found in some of the later herbals. Camerarius also issued his *Hortus Medicus et Philosophicus* at Frankfort in 1583. The cuts in this book may also have been part of Gesner's store. Others, again, were not printed till they were published with other posthumous work of Gesner in 1753 at Strasburg.

Gesner's cuts are more botanical than any others of the time. He much more carefully noted the details of the plants, and gave on a larger scale various parts and dissections. These figures are just beginning to lose the qualities which make them useful to the artist. The more botany, the less there is of the individual plant.

In England two noteworthy books were published by John Parkinson, one of the most beloved names in the history of botany in our country. "Parkinson's Park," or, in its own punning title, *Paradisi in Sole Paradisus Terrestris*, was issued in 1629. It has recently been reproduced. It contains a great array of "pleasant flowers" and kitchen-garden plants. The cuts are largely re-engraved, very crudely, from Lobel, Dodonæus and Clusius, and particularly from the copperplates of Crispin de Passe—*Hortus Floridus*, 1614. Some may also be from Besler's *Hortus Eystettensis*, issued at Nuremberg in 1613, but I have not compared the two. Parkinson's other work is his *Theatrum Botanicum*, London, 1640, in which the figures are also poor copies for the most part of those of former writers.

Such were the different schools of plant-drawing which have made the word "herbal" notable. Herbals as such continued to be issued up to our own time, but the illustrations in them are of no consequence.

As the seventeenth century opened favour turned towards copperplate engraving and towards etching. The copperplate frontispiece by William Rogers to Gerarde's Herbal, 1597, exhibits a number of plants very well expressed, but the difficulty and inconvenience of printing retarded the general introduction of the process. But in 1613 Besler's unwieldy *Hortus Eystettensis* was issued at Nuremberg. Its figures are "full size" copperplates, and are not of great merit. They have about them that peculiar conventionality of rendering which the process readily fosters.

Of much higher order were the etchings with which Fabius Columna illustrated his *Phytobasanos* and his *Ecphrasis*. Columna was a competent artist and also a learned botanist. A number of reproductions from the *Ecphrasis* are printed in the Appendix. In them we see united truth and beauty, and they are indeed as far from the conventionalities of Besler as they well could be. Particular attention may be drawn to the leaves of the specimens of Hieracium.

In the Appendix are placed some engravings from the *Hortus Floridus* of Crispin de Passe. This rare little book was issued 1614–1616, and it may be questioned whether better copperplate figures of plants were ever executed. They differ vastly from the drawings of Fuchsius, and were done from a different point of view. The number of plates in Crispin is small, and those of the *altera pars* are very poor, and certainly by some one else.

Whether any *good* copperplates were done afterwards in the long years which intervened before Sowerby and Sydenham Edwards began working for Curtis's *Flora Londinensis* and *Botanical Magazine*, or for their own *English Botany* and *Botanical Register*, I do not know. But the rise of strict botanical science, demanding the diagram rather than the figure, put an end to the production of such drawings as it has been our task in this volume to collect together.

Great Britain seems to have contributed little or nothing to the illustration of the famous herbals. We hardly dare claim that the figures in the *Stirpium adversaria nova* of Lobel and Pena, issued at London in 1570–71, were of English origin. More ground there may be for adopting the crude cuts which appeared in a very rare little work of only about fifty leaves— *La Clef des champs pour trouver plusieurs Animaux, etc.*, by Jacques le Moyne, published in London in 1586. Of these cuts a few specimens are given in the Appendix. Their "decorative" qualities must be assigned to the same origin as that which rendered the figures in the herbals ornamental —an economical use of the area of the block, a bold and certain manner of drawing, a regard for the obvious essentials, and an absence of fear of botanical inaccuracy. But what in these crude cuts most arrests attention is the definite likeness of some of them to the figures in the *altera pars* of the *Hortus Floridus* mentioned above. That some of them had a common origin there can be no doubt.

Perhaps the earliest English work of which we can be sure is the title-page to Gerarde's Herbal, 1597, by William Rogers, already referred to. It includes numerous flowers and plants, rendered so well that we wish there were definite figures of plants by him.

CHAPTER II

THE USE OF PLANTS AS ELEMENTS IN DESIGN

SOME people regard Nature as a great storehouse of Form, and they rifle it with predatory hands—making no acknowledgment. They use the forms of natural objects without caring at all what the objects are. The outline of a spider's leg gives them an " idea " for the handle of a jug—and so on. Whether this is a legitimate way to use Nature or not, it is not the intention of the author to discuss in this chapter. He would indeed rather let the question alone as one which settles itself. For the success of a piece of design is based upon the observance of a spirit of calm, architectural fitness, and so long as that is present it matters very little by what means the artist arrives at his result.

But it is not likely that Nature used in that way will yield that which is most enjoyable.

All design is synthesis—composition—the putting of forms together so that they conform to a demand which is made by the eye and the mind. Why this demand is made by the eye and the mind it will not profit us to inquire. But the demand asserts itself at every turn in life's affairs. It is based upon some adoration of Order amongst Variety.

All people arrange things in orderly fashion. To do so is one of the most general of human actions. People can therefore be trusted, when they arrange, to arrange orderly.

Design is conscious arrangement. Spite of the objection that is sure to be raised by those critics who, not being producers, delight to find some paradoxical *impasse* to block the efforts of others—spite of their complaint that art must be artless, one repeats, what every productive artist knows, that design is conscious arrangement.

Let it not be thought, however, that designing, any more than walking, should be done with a mental calculation, laboriously conducted for each step. The conscious control of a man's powers in walking has become by usage so incorporated in his being that it almost deserves another name. In designing also, one designs with little success till the powers one uses have become automatic in one, and act of themselves, as it were.

Nevertheless, the early stages in designing must be as painfully conscious and laborious as are the child's efforts to walk.

What it is which guides our choice in the formality of arrangement, we need not, as has been said, inquire. It is sufficient for us to review those peculiarities of arrangement which we come to call the laws of design. These laws we find out by analyzing those works of art which please us, and by watching the processes which our faculties adopt when we are ourselves at work.

We find that the great laws of design are geometrical, and do not arise from imitation, or what may be called portraiture.

In a piece of pure portraiture it matters not at all what disposition the masses or lines of the drawing make, though some people think that it does, and that some care should be taken to make the work pleasant to the eye. But in decorative art what is in portraiture demanded by some only, is demanded by all, except only those who overlook the whole effect and being of an object, and fix their eyes upon some detail or circumstance connected with it. The fact that our daughter has done this or that leads us to engross our attention upon parental sentiment. A favourite yellow, or blue, will blind some people to the most harrowing defects of form and design. Skilful workmanship will, with some, outweigh all æsthetic considerations, costliness with others, age with others, and mere weight and size in many more. All these are instances of a sort of aberration which seizes upon the sanest of mankind when works of art are in question.

But these instances of misjudgment do not affect the main belief in order amongst varied things, which is one of the predominant characteristics of human consciousness.

This universal belief in orderliness comes back in artistic matters to geometry, that is, to measurement—to measurement of distances, areas, angles between lines, and of the chords of curves—to use but a few simple expressions.

Now geometry, as one commonly meets it, is a simple affair. One stops short in one's explorations in it wherever one pleases, and the impression left on most people's minds is, that if one knows a certain amount of the simpler part of it one knows enough for the ordinary affairs of life. But geometry is a never-ending science. The relation between measurements can be carried to any extent. Factor after factor can be introduced, and the most complicated problems assailed. Now the designer is in a peculiar position as regards geometry. For the actual practical part of his work he needs very little of it. But of the sense of the relationship of measurements he can never have too much. Design is based on measurement and on geometry. True, but it is based not only on the simple geometry which we use to set out the repeats of our patterns, but on that extremely abstruse geometry which must be felt rather than learned.

It is for this reason that it is impossible to design by geometry, though design is based upon it. It is easy enough to make some sort of pattern by

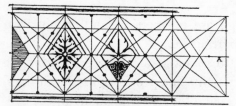 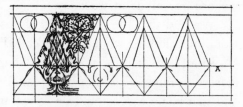

III.—A PATTERN SET OUT UPON OBVIOUS GEOMETRICAL LINES.

IV.—A PATTERN SET OUT UPON LESS OBVIOUS GEOMETRICAL LINES.

geometry, but what of the subtle variations we have to introduce before the pattern pleases us! They are apparently subject to some remote laws of arrangement, for we are conscious that they hang together in some system, and are not haphazard. But to find any rule, or to apply a rule if found, carries us at once beyond the bounds of practical artistic work.

Consider in support of this the two designs here given. In the first, the heart of the design is carried on a line, A, passing through the middle. It is easy to see how, when this is the case, many measurements fall equal, or are in simple proportion. But when, as in the second, the line A is dropped a little lower, the adjustment of the proportions becomes a matter far beyond any simple calculation. In both patterns there is a triangular form, with a smaller triangle within it. In the first pattern these triangles are proportionate. They are within parallel lines; but in the second the inner triangle cannot be of the same proportion as the outer. The slope of its sides is determined by the taste of the designer, which is another way of saying by his sense of the geometrical relationship which is secretly governing the parts.

But the harmonious arranging of parts is sometimes of a very obvious character. Thus mere parallelism, as A, Fig. V, is a positive and most valuable form of harmony. The same applies to parallelism of bent lines, B. In C the spaces are regularly diminishing. In D curves are regularly changing; in E sizes; in F and G lines are harmonized by radiation, whether they continue to the point of radiation or not. In H there is the principle of the steelyard—balance— with the leaf as fulcrum, and the larger form nearer to it. Such are the obvious means of harmonization, and must be consciously mastered before a faculty of designing has taken the slightest root.

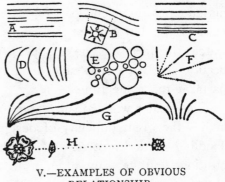

V.—EXAMPLES OF OBVIOUS RELATIONSHIP.

The next essential power in the designer is that of securing architectural stability. This means mastering and appreciating what I have elsewhere called the architectural and the sub-architectural lines. These are the horizontal and vertical lines. Those which are sub-architectural in their character, that is, partaking of the nature and stability of architectural lines, and yet dependent upon them, are the circle, the pyramid, and the Gothic form of arch, which is a compromise between the other two. The spiral curve is sub-architectural, and it will be found that it always excuses itself. That is to say that a stem, if of a sub-architectural character, will seem to be so much of the kind that it should be in a work, that its not being the natural curve of a plant will not give offence.

What we commonly call a geometrical setting is really a network of architectural, or more commonly of sub-architectural, lines. Whether composed of circles, pointed ovals, hexagons, or any other form of that character, the pattern will hold together rigidly, and cannot but be satisfactory in effect.

VI.—ARCHITECTURAL AND SUB-ARCHITECTURAL LINES.

But the application of the principles of arrangement has to go so far that not only is arrangement achieved, but beauty also. In beautiful forms there is to be noticed a just correlation of parts precisely of the kind we have been considering. But so subtle is this combination that no law or rule can be made to produce it. No rules can take the place of judgment. So long, indeed, as the mind is concerned with rules, and the application of them, so long are the faculties only partially active. It seems that in artistic work all the faculties of a man must be in play at the same time. He must not overlook any considerations, and must bear in mind all possible contingencies. Nothing is easier to the conscientious person than to put his energies upon one part of his work at a time. To do so appears very businesslike, but it is not the process of art. The designer must therefore watch the forms as he produces them. He must watch their growing beauty, and, with eyes wide open, must feel the effect and value of all the parts at the same time.

The reader will probably have gathered from what has been said wherein lies the great difficulty of designing. One has to train oneself in the handling of the most obvious kinds of arrangement and geometrical disposition, and yet any attempt to make a design by relying upon these obvious kinds of arrangement will usually end in one's producing an exceedingly dull, mechanical, and uninteresting pattern. While we know by analysis that what we call principles are enduring forms with that subtle correspondence of parts which is the chief mark of beauty, we find the exercise of these principles helps us but little ; and yet we dare not neglect them. We have to trust to our own choice after all, and end with the simple belief that what pleases us is beautiful. Indeed, no other rule is of any use to us, and if we do but honestly please ourselves, and make forms which genuinely give us pleasure, we shall find ourselves credited with a power of designing beautiful things.

The elements of the design—its details—must be as carefully formed according to the principles of design as are the main lines of it. And here we must recognize that we must throw aside all fear of offending the botanists, or even of offending the gardeners. Much as we wish to base our work on identifiable images of plants, we must, above all, remember that we are decorators first. The old drawings appeal to us as so suggestive for design, because there was, in the days when they were made, no fear of botanical inaccuracy. Now we are so fearful of a false rendering, and so afraid of a bold free line, that our designs lack ease and confidence. But if one looks at the designs of William Morris, one sees how little he was abashed by any taunt of the scientist. Nor, again, was he afraid to swing in his lines boldly. In one design he has lilies with five instead of six segments to the perianth.

One cannot proceed far in designing if one has not the mastery of some elements to one's credit. For, although the success of designs is built upon the arrangement and proportion of great masses, it can be marred, or even ruined, by ill-formed elements. Moreover, it is extremely unlikely that one who makes poor elements will make good designs, for an element is a little design itself.

The designer must therefore first prepare his elements. He will soon find how great a variety of forms his attempt to use nature will give him.

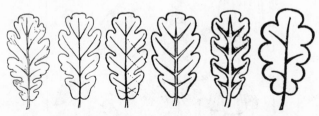

VII.—THE OAK-LEAF AS AN ELEMENT IN DESIGN.

Some examples are given here.   In Fig. VII, which represents the oak-leaf, it will be noticed that the leaf upon the left is a natural form, and those towards the right depart more and more from it.   This departure is an ornamentalizing of it, and is sometimes called "conventionalizing."   This term is not a bad one, but the practice is often not commendable, for instead of adapting the form to the use pressed upon him, the designer merely gives the form a new appearance, altering its curves and shapes, and generally trying to convert it into one of the worn-out elements which are flying in the air at the moment.   Conventionalization is really the modification of the form because of certain conditions under which the form has to be used.   The convention varies with every piece of work one undertakes,

VIII.—EFFECT OF DIFFERENT KINDS OF VENATION ON AN ELEMENT.

and one cannot conventionalize a form once and for ever, but must treat every case as it arises.

For after all, decorative art, where it introduces identifiable form, is practising portraiture—only, to make the portrait "tell," we must adapt the rendering to the conditions.

Variations in conventions are due, of course, to differences of mood. Some artists work habitually in a stern mood, others in a gayer, and those of the former class will drive their nature-forms into a severer mould than the others.

IX.—ELEMENTS DERIVED FROM WOODCUT ORNAMENTS. C. 1630.

X.—ELEMENTS DERIVED FROM WOODCUT ORNAMENTS. C. 1630.

In Chapter III is given some account of Jacobean Floral Ornament. Here, in Figs. IX and X, are some elements derived from ornaments there referred to, and reproduced. They will be seen to be largely identifiable.

In Fig. XI are given elements derived from figures in the body of this book, showing the use that can be made of them. The elements are Sloe-flowers, a Bramble-flower, and a Currant-leaf.

The elements being prepared, the next task is to set out the pattern. Perhaps the simplest setting is that given in Fig. XII. Within the squares formed by the vertical and horizontal lines some sort of foliage, some elements, have to be arranged in order to fill them completely. Here they are filled by a flower—what flower? it may be a rose, a poppy, or a crow-foot—since it has so many stamens. If its petals are to be counted it must be called a poppy. To reduce petals from five to four is no crime in decorative art.

From such a rigid and simple planning as Fig. XII one can depart towards a looser and freer arrangement, but the main setting must be sub-architectural in character. Within the network of hexagons, squares,

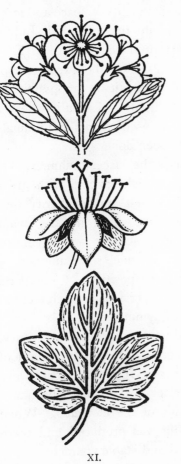

XI.

XII.—A PATTERN SET OUT ON STRICTLY
ARCHITECTURAL LINES.

scales, or whatever forms are employed, will come other forms also of a sub-architectural character—flowers forming circles or squares or pentagons, stems of spiral curvature, and twigs spreading out fan-wise. Mingled subtly with the roundish lines, and the smooth or tangential radiation, will be angular lines and radiation of a sudden character. It will be found that designing comes down ultimately to space-filling, for unless the spaces are properly filled they are destroyed, lose their shape and are not what they are supposed to be. Indeed, good space-filling preserves good spacing, and without it good spacing is of no avail.

XIII.—A PATTERN BUILT UPON A HEXAGONAL NETWORK.

From such a pattern as that in Fig. XIII we may get to others of still freer structure, and finally to such delicate adjustments of space and form as is seen in the beautiful Indian drawing reproduced in Fig. XIV (*Frontispiece*). Of the excellence of such a drawing it is impossible to speak too highly. I doubt whether better drawings of plants and animals have ever been done.

We must not forget how the Japanese arrange their freer ornament. It is by a system of balance carried out with great delicacy and with great variety of proportion. It is in the great variety that its peculiarity lies. There are European instances of balances being used as the main principle in design—notably in illuminated manuscripts, and in much of the work of the eighteenth century, but, as a rule, some method of arrangement, and some predilection in the size, or position, of the parts becomes manifest. It is indeed the constant reversion to architectural conditions which keeps introducing some form of symmetry.

Apparently the error which was committed in the eighteenth century, and in the nineteenth up to the time of Morris, was the repetition of unsymmetrical forms, such as bunches of flowers. It is an error always made by people whose grasp of architecture is slackened. The only chance of success in such work lay in the dense massing of the elements. When this was done an oval or round form generally was produced, and these ovals or circles took naturally a sub-architectural and stable character.

XIV.—IN THE JUNGLE.

An Indian drawing, reduced to about two-thirds the scale.   Executed about 1570.

[*To face p.* 18.

A HEAD-PIECE USED BY STEPHENS, OF PARIS

CHAPTER III

JACOBEAN FLORAL ORNAMENT

OWEVER clumsy Jacobean ornament may be, it is, if not admired, at least loved. It belongs to a style which, however many parallels to it there may be in Continental art, seems to us peculiarly English. The clumsiness, perhaps, deceives us, and fosters a conception in our minds that there was in those days more of British bungling-through than of skill or principle. A closer acquaintance shows us that this was not the case. It is much to be doubted whether what we call degenerate styles are not as much the groping after new principles and new ideas as of the misuse and abuse of something inherited and not understood. Give the degenerate style time and opportunity, and it will discard the forms it has outgrown and does not know how to wear ; for, unless thought and civilization are gone, they will express themselves again in their new conditions.

To reject skilful methods, to neglect traditions, is not necessarily to be either less civilized, or less thoughtful, or less devoted to art. It may be that the old highly developed methods and modes have lost their meaning —have become highly organized craftsmanship, to the neglect of that impulse and meaning which called them into being. Indeed, it is very clear that this was the case with the ornament of the later sixteenth century. No healthy human instincts and passions could longer have endured the acanthus foliage and cartouche forms of that time.

Acanthus foliage and cartouches constituted the official ornamentation still in vogue when the seventeenth century opened. It was essentially of a modelled character. Arising in carving, it became so generally used as to find a place in practically all the crafts. Certainly where there could be shading with which to simulate a carved effect it gained a definite foothold.

Carved ornaments are usually distinguished by pure and decided curves. The spirals, for instance, are very regular ; and the same regularity is to be seen in the rosettes. It is evident that this regularity can be adopted in carved work, because the light and shade at once vary it. The regular spiral is no longer regular, nor the rosette strictly symmetrical.

But it is the fate of all highly trained and sensitive art to suffer in essentials, as soon as a less skilful hand touches it. The blunter forms of art are readily repeated and lose little in inferior hands ; but sometimes the greatest skill is needed, or the work degenerates at once.

One of the best instances of this change is in the re-engraved botanical drawings in the herbals. The less sensitive suffer very little, the more sensitive lose all their beauty. The acanthus and cartouche style with which the sixteenth century closed was of that sensitive character which is at once destroyed when wrought indifferently. The change can be easily followed in the woodcut book decorations of the time. The acanthus and cartouche style required not only skill but understanding. It was purely artificial. It responded to no natural instincts or ideas. Its forms were not the forms of actual life, but pure make-ups ; and make-ups that meant nothing at all, not even to their skilled producers, much less to the unskilful workmen who endeavoured to copy them.

The inevitable happened. The more rugged artists, lacking the necessary education to assimilate and emulate the examples set before them, were yet artists. They were engaged in making things the better by decoration, and found in themselves principles and modes which, if not leading to the refined and artificial art of other times and other places, were full as trustworthy as guides in design.

It will be found to be true that there are two main classes of ornamentation, and that the one succeeds the other, like the swing of a pendulum, when the one which is in vogue declines. An age pursuing good technique will come to revere a kind of work in which technique is lavishly displayed, till at last the work depends entirely upon the technique. That is to say, the mental suggestion, the imitation of nature, which at first were recorded in that technique, become neglected as skill becomes more and more insisted upon. When, therefore, a style buoyed up by skill (as in that of the Roman and Renaissance) is suddenly bereft of skill, the interest which has come to be wholly in the skill is lost, and the artist reverts to the imitation of nature and the expression of tangible ideas. We see this in the Byzantine and in the Jacobean.

Further, skill implies form and drawing more than colour—colour may need skill to produce it, but it rather needs feeling and passion. Form and line, on the other hand, must be rendered by skill. So that a skilful style is usually one of form and drawing, and, when it declines, it gives place to a style based upon colour, or upon black upon white, or white upon black.

When, therefore, the seventeenth century opened in England, the acanthus foliage and the cartouches were still in vogue, but were already suffering decline; while at the same time the flower-forms, which, as the emblems of England and Scotland, could be appropriately used, were introduced more frequently than was actually necessary. Gradually the floral elements became more and more numerous; and the acanthus and the cartouches sank more and more into insignificance.

The acanthus was, however, by no means discarded. In a very blunt, coarse, and rather flat condition it remained a most valuable ornament. It could be repeated without offence, and without much notice being taken of it. It was used, in fact, wherever a more severe ornament was advisable, generally closely connected with mouldings and strapwork. It occurs in needlework borders wherever a steady repeat is needed to render the design architectural and formal.

The strapwork was indeed no more than the cartouche of earlier days. The cartouches, when in their glory, had been boldly curved in form, so that they took definite light and shade—one side being in light, the other side in shade. Twistings, piercings, and involved convolutions had been employed to develop the form as much as possible; and with the form, the light and shade also were consciously elaborated and carefully balanced.

But as the decline set in, the surface of the cartouche became again flat, and only the edges turned or curled up. The piercings still remained, and indeed were developed, for there was no other way of using flat cartouche work than by piercing holes in it, and letting it make a network pattern, dark, upon a white ground. Thus arose rather gridiron-like designs, familiar enough to us; and these continued, growing flimsy and slight, and ultimately, becoming mere lines, as the seventeenth century wore away.

The rose figures largely in Elizabeth's day, and to it is joined the thistle when James succeeds her. Designs including the rose and thistle are thenceforward common. The lily of France, very poorly represented like a fleur-de-lys on a horse's harness, sometimes appears as well.

Although 1630 is perhaps not too late a date for the rise of the general use of floral patterns, such really arose in the days of the Tudors, and of course floral forms had been used on textiles for a hundred years before that. About Elizabeth's time, however, there arose a crude, but beautiful, kind of decoration in needlework (of which examples are at South Kensington) unlike the official work of the day. It consists of circular curves of uniform size unconnected—and hardly complete—and each enclosing a flower. The flowers differ, but recur without any system.

It is precisely the kind of work a person too unsophisticated to know anything about acanthus foliage or the complications of organized design might well do if called upon to decorate some surface. Possibly the great increase in the study of plants, which at that time was promoted by the publication of profusely illustrated herbals, and the great impulse then given to gardening, directed attention to the forms of different flowers.

There, nevertheless, the patterns are, combining severe scale-like planning with a simple rendering of nature. The flowers most generally seen are the rose, the pink, the honeysuckle, the columbine and the thistle, with numerous others less easily identified.

The style seems to have suddenly expanded in the reign of Charles I. It does not seem to have lingered long after the Restoration. The acanthus once more came back, this time with more punctilious execution, and with a promise of the boisterous rococo to follow. The classic taste revived, and one no longer sees the roses and tulips, the oak-leaves and columbines of earlier days.

This floral style has been the subject of both admiration and dislike. It has about it a peculiar " old England " flavour, which has prompted that fervent appreciation due to sentiment and patriotism. On the other hand, its monstrosities have called down upon it the censure of the refined and learned. Its complete disregard of scale (with a tiny lion and an elephantine fly side by side with a huge pansy and a diminutive cottage) has been chiefly to blame for this. Its, at times, gross ignorance of the correct forms of conventional ornamental elements, as of acanthus leaves, earns for it the distinction of being rather a block-head style; and—there is no gain-saying the fact—at book-learning its producers must have been uncommonly dull.

Nevertheless, the producers of this style, if clownish, were by no means so stupid as decorative artists as they appear to be at first sight. They achieved results which are rich, decidedly pleasant, and thoroughly enjoyable. It may be the gardening, the flute-playing, maypole-dancing part of us which is delighted: but it cannot be denied that it is the more natural and English part of us.

The style is peculiar for its use of many flowers on one stalk. This is, by some, regarded as a wicked abuse of Nature. But, in a garden, still less in a hedgerow, one does not see the various plants clearly distinguished from one another. They raise their heads up amongst their neighbours' foliage, and intertwine their boughs, without regard to the preservation of identity. Such a " natural " confusion may be, perhaps, a sufficient answer to the charge of violating Nature's methods.

Indeed, a love of Nature is the keynote of this style. There is in it none of that ornamental twisting of plants which so prevails at the present time. There is the bold use of the spiral main line as a necessary decorative member, but beside that there is only the straightforward representation of the plant.

When the style arose the identity of a plant was confined to a few bold characteristics, generally concerned with the main form of the flower and of the leaf, although the whole appearance of the plant and the form of its root were always noted. Still, so far as the ordinary delineation of plants was concerned, a flower, perhaps also a bud and one or two leaves, sufficed. Possibly the habit of gathering flowers and of putting them in baskets and bowls may have had something to do with this pre-eminence of the sprig and the neglect of the plant as a whole. Or it may have been the not un-natural hesitation of those unskilled artists to repeat the elaborate, fully rendered plants which must then have been so commonly to be seen on the tapestries hanging in every large house.

To whatever cause we ascribe the prevalence of the sprig as a unit of decoration, it became an important detail in all ornamentation. It partakes of a very decided character. It has usually a single- or double-curved stem with a large flower at top, while from the lower end branches off a leaf which crosses the stem. The decorative quality of such an element will be at once evident to a designer, who will recognize in it that reversal of line which renders the form suitable to stand unenclosed.

One is, therefore, not surprised to find this crossing of leaf over stem in those early patterns in which little sprigs occur within scale-like, spiral, or circular curves. These curves were all of the same size, and in the early examples were not connected. In the later patterns they are connected together, and grow from one root. And although there are many small spirals, enclosing each its separate flower, there is at the same time, in obedience to the demand for decorative subordination, prominence given to several large, bold curves which subdivide the area treated.

These bold curves are, as it were, the branches of this all-bearing tree.

XV.—JACOBEAN PATTERN, ILLUSTRATING THE SPRIG WITH A DOUBLE CURVED STEM AND THE LEAF CROSSING IT.

XVI.—GILT POTTERY ORNAMENT. THE LAST TRADITIONAL REMAINS OF THE JACOBEAN STYLE.

The tree itself has an impartial character. It usually spri gs from a root, or nest, of clumsy acanthus leaves, which are unidentifiable as representing any plant at all, and the same conventional foliage is usually further called into requisition at the junctions where the " branches " divide. These pieces of crude conventional foliage are apparently of great value in steadying the design. They are usually of considerable area, and seem to take away the fussiness which the multiplicity of natural forms might otherwise produce.

In the upright wall patterns (or, more properly, curtain patterns) the same conventional foliage is used. In these patterns broad stems arise from a conventionally rendered ground below and, in wavy lines, stretch up to a similar border above, throwing off large conventional leaves which serve in this case the same purpose of checking fussiness. We may indeed note of these patterns that the largest forms are not natural, and, also, that the pattern is not based on a geometrical planning, but is formed in obedience to a sense of growth.

In setting out their patterns, the designers seem to have been careful to boldly place their large masses in those positions which architectural

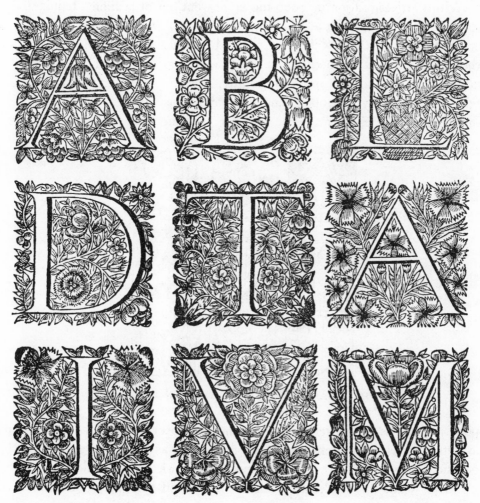

XVII.—INITIAL LETTERS USED IN THE "SEALED BOOK" OF CHARLES II, A.D. 1661.

considerations demand. In small panels they place a mass in either upper corner, and a heavy mass in the middle below. They very carefully place the rounder parts of their main lines as if those parts were of particular importance. They seem, after these details were properly placed, to have cared little for that perfect tangentiation of which the artists of acanthus design at an earlier time were so considerate.

Still, there is a peculiar beauty in the main lines and in the details, for which it is hard to account. The curves have a variety not readily recognized, and never lapse into designer's trills as they invariably did at a later date.

With the Restoration in 1660 the style went out of fashion. It lingered on, and still lingers on, in a debased form. The flowers became all starry, or rosettes; the curves all graceful and skilful; identity of plant was wholly lost; leaves all became, as did the flowers, alike. The basket of flowers, at first so formal and so decorative, became irregular, tilted on one side, suspended by a ribbon, and ultimately has fallen so low as to be shunned by all designers. The jeweller still engraves the flower of five

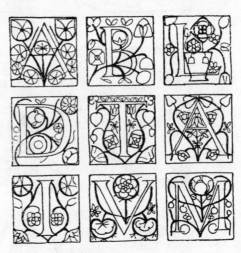

XVIII.—ANALYSIS OF THE INITIALS IN FIG. XVII.

petals, with its nondescript leaves, and its rococo cartouche work, on lockets, spoons and watches, but the style as it was has gone, save for such occasional revivals as are seen in good plasterwork and embroidery.

Even while this work has been in preparation the reversion to this old English style has become marked. Professor Lethaby has definitely encouraged the study of it; and examples are now freely reproduced. The student who begins to look for the floral pattern will indeed find it at every turn. At the Victoria and Albert Museum are many most interesting specimens.

It is a little out of our way to inquire into the designing in detail of these patterns. Some examples of Jacobean printers' decorations are used as head-pieces and tail-pieces to the chapters. That to the Preface is not an old example.   The initials used in the sealed Prayer Book of Charles II are extremely handsome.   Some are reproduced in Fig. XVII, and are analyzed in Fig. XVIII.

What, however, above all needs to be impressed upon the designers of to-day is that they must not merely learn to imitate, or rather "forge" past styles, but having learnt from them, must work more hand in hand with Nature than is often the case.

CHAPTER IV

SOME CONSIDERATIONS GOVERNING THE USE OF PLANTS IN DESIGN

THE designer must be master of conventional design—design, that is, which uses non-identifiable elements. Not nondescript elements, for they are the last to be desired, but elements of such simplicity, or of such an architectural character, that they may be said to be the products merely of the play of decorative principles. Such an ornament is the Greek Honeysuckle. Such an ornament is the Acanthus. These, we may be told, are imitations of nature. But whatever their origin may have been, they were used not as imitations—portraits—but as decorative elements of a kind more closely connected with architecture than with plant-life. That in such ornaments there is life and growth does not declare them imitative of natural form, but only demonstrates that in conventional form there is a power of expressing qualities which we readily associate with, and regard as indicative of, plant-life and animal-life.

Of the historic styles which fulfil the requirements of being useful conventional styles, the Early English Gothic, and the Byzantine or Romanesque are perhaps the safest to build upon.

When the designer can decorate by using some such style in which there is no added difficulty of portraiture, he should seek to vitalize it by a constant study of plant- and animal-forms. Without trying to weave into his work any of them he will gain in hand and eye a sensitiveness to the vigorous and interesting rendering of form which will raise his productions above a diagrammatic or exercise-like condition.

There should be no hurry to introduce actual plants, and certainly no hurry to include plants amongst more conventional forms. Indeed, as he grows in sensitiveness, the designer should try to get all he can out of mere geometrical forms. The geometrical forms will then become endowed with subtleties which will give them that peculiar quality we call beauty.

When this stage has been reached the designer should begin to strew plants and flowers over his work—not to twirl and twist some unfortunate Cornflower or Poppy into a triangle, or a circle, or a square, with the squirmy lines dear to the student and trade-designer of the day, but letting floral-forms and plant-forms grow into his patterns if they wish to, and if he cannot keep them out. Or he may frankly portray his plant according to the necessary limitations of position and process.

Position and process will compel some simplification to be adopted—probably some formalness of arrangement, perhaps symmetry—perhaps an enlargement of the flowers in relation to the other parts of the plant, perhaps a unification, or a variegation of colour.

The chief errors in floral design are usually two. Sometimes the plant is not really cared about at all, and is merely adopted because it gives some variety or novelty to the work, and is supposed to afford suggestion. As a matter of fact, in the hands of any but gifted designers it usually loses all its natural character and charm, and is bad as ornament—bad probably because, maltreated as it is, it is preventing a simple decorative solution of the problem.

Sometimes, on the other hand, the plant is a mere still-life rendering, lacking all decorative qualities, and is looked at merely as a painting, and not as decoration. Much needlework is of that order. Usually, it must be admitted, the people who work these designs have some regard for nature, which the former delinquents have not, but they have, as a rule, neither a power of drawing nor of true colouring, and are shockingly deficient in a sense of decorative fitness.

It is doubtful, however, whether in this last respect they are not in better case than the producers of the "new art" monstrosities, who seem in art analogous to the well-educated amongst criminals. All they know is used in the perversion of good things.

Let the designer recognize at the beginning that he is not obliged to use plants at all, that it may be his lot in life not to use plants ; and that beautiful objects could be produced without the use of them. And, indeed, until the designer adopts that attitude, he will misuse his plants. Until he can do without them he cannot properly introduce them.

Assuming, however, that the designer is fit to use plants—and he may show that he is in his first design, if that mysterious development of taste and feeling has gone on which sometimes does go on without any positive cultivation—he must combine the architectural and structural conditions with a portrait of the plant. That peculiar balance of mind which can determine the precise quantity of each quality—naturalness, formality, simplification— is the most desirable of gifts in artists. It is the gift of feeling ; and by no rules can the determinations which are made by it be arrived at.

All that we know is that to the grasp of rhythm has to be added the real positive love of the plant. This love of the plant will correct flatness ; it will drive away insipidity, it will be a source of inspiration. But all our love of nature—of the plant—will be thrown away if we forget the great stress of rhythm and architectural fitness without which our plant-portraits will dwindle into practically invisible, though probably highly commendable, studies of plant-form.

People forget how much the Japanese print owes to the violation of scale which the artist permits himself. People forget that many of the plants in Japanese drawings are " artist's plants."

In portraying our plants we must then permit ourselves a bold hand, but must not forget our plant. Its character must be preserved, but its character does not reside wholly in its small angularities, and other minor peculiarities as one is so easily tempted to think. Yet no one can leave out these peculiarities and details unless he can put them in !

What, in a word, we need, is some measure of the mood of Kate Greenaway. Perhaps no examples are better worth studying than the flowers and floral designs with which she decorated her books.

CHAPTER V

THE GENERAL FORM OF PLANTS

VERY little observation shows us that plants are, broadly speaking, of four kinds. Some are without true leaves and stems —the Fungi and Lichens. Others are leafy but lack flowers— the Ferns and Mosses. Others have parallel-veined leaves, and their flowers in triple formation—the Lilies, Hyacinths, Grasses, etc. And others have spreading leaves, flowers in quadruple or quintuple formation, and have their timber in concentric rings—the Buttercups, Roses, Oaks, etc.

In the next chapter the botanical classification of the vegetable Kingdom is given, and it will be seen that the distinguishing marks by which the different "classes" are known are of importance to those who represent plants in drawings or in designs.

Here we are concerned to go over the various parts, or members of plants, in detail.

When an acorn begins to grow it sends down a root, and sends up a stem, while the main bulk of it divides into two pieces which are called cotyledons or seed-leaves. Plants whose seeds produce two cotyledons are called Dicotyledons. They are also called Exogens, because they increase at the outside, as is seen most pronouncedly in the "rings" of timber when a trunk is cut across, where the youngest wood is at the outside.

Exogens are the most highly organized of plants. They have more details, more parts with specially assigned functions, than any others.

Plants whose seeds produce but one cotyledon are called Monocotyledons. They are also called Endogens, because they increase within. Their timber is not in rings, but is of confused structure, with the youngest wood in the middle.

These two great classes, the Exogens and the Endogens, embrace all the flowering plants, leaving outside the Cryptogams, which have no true flowers—the ferns, mosses, fungi, lichens and sea-weeds.

The exogens and endogens are not wholly dissimilar. Indeed most people, even those who are fond of gardening, have usually not observed that the Rose, the Buttercup and the Hawthorn differ in certain well-marked characteristics from the Lily, the Iris and the Grasses.

In Europe all the larger plants, the shrubs and the trees, are exogens, but in tropical countries the endogens assume great size—as the Palms, the Dragon-tree, and the Banana-tree. Such an un-herb-like substance as Vegetable Ivory is the produce of an endogen.

But with us endogens are nothing more than herbs. They provide us with the most valuable of vegetable produce—wheat, and with some of the most beautiful of our flowers—the lilies.

The essential organs of all flowers are (1) an Ovary, in which the seed is matured ; (2) a Stigma surmounting the ovary, or perched upon a Style, for receiving the pollen by which the ovules in the ovary are fertilized ; and (3) a Stamen which produces the pollen from two pollen-sacks upon it. Let it be noted here that the ovary is nothing more than one or more leaves endowed with the property of producing at its edges a number of ovules or immature individuals. Cut a capsule of the columbine across, or split it open, and this will be seen to be the case. The Anther, or head of the stamen, is simply a leaf which also produces minute individuals on its edges. It is the union of those in the ovary and those out of the pollen-sacks, through the agency of the stigma, which produces fertilization and living reproductions of the plant. Compare in this matter the spore-cases in ferns, and particularly in such a plant as Ophioglossum.

The essential organs are therefore an ovary crowned by a stigma, usually raised upon a style, and a stamen. The first " class " in the Linnæan system of classification is Monandria—having one stamen, and its first " order " is Monogynia—having one pistil. Among British plants fulfilling these simple conditions is Centranthus ruber, the Red-spur Valerian.

When the ovary is simple, the stigma is also simple, and formless, but not necessarily. When the stigma is lobed or divided it generally indicates that the ovary is in several cells, or lobes, owing to its being formed of two or more ovaries conjoined. Hence in Geranium we see five stigmas at the top of the style, and the fruit develops ultimately into five separate pieces. Now it is very common for ovaries to be thus made up of several—the apple contains five, the orange more, as any one who cuts these fruits across can see. Therefore if there is but one style, it will generally be found lobed or subdivided. How many divisions it has is important for the determination of its place in the Natural Orders—as that in the Daisy-order it is two-lobed. Sometimes, although the fruit is a consolidation of several cells, it does not follow that their styles are united into one.

For the draughtsman, indeed, it does not matter very much what the relationship is between the stigmas, styles and the fruit, but he has to note the number of them, and he may remember that where there are separate or many styles, as in Hellebore, these styles will probably not be much thickened at the end, and, I think, never divided; whereas, when there is but one it will generally be thickened, and lobed or divided.

The stigmatic surface is generally glistening, and covered with some sticky substance.

Exogens and endogens do not differ in their styles and stigmas, and although endogens have the parts of their flowers usually in threes, the stigmas are not always three. In grasses there are two feathery styles. In Lily the style is thickened above and three-lobed. In Iris the style spreads into three petal-like portions, or stigmas, a small part of the under surface of which is stigmatic.

There is little difference, again, in the form of the stamens. A stamen consists of a filament or stalk supporting an anther. The filament may be slender and thread-like, or it may be broad and flat like a leaf or petal. It is usually colourless and pellucid. The anther consists of two pollen-sacks, which, if on either side of the filament, are said to be *adnate* to it. If the pollen-sacks are side by side and are perched upon the filament in continuation of it, the anther is said to be *innate*. If joined by its middle to the summit of the filament hammer-wise it is said to be *versatile*. The filaments are sometimes very short, sometimes connected together below; sometimes they rise from the ovary, or from beneath it, or from the calyx or from the corolla. In a few cases, chiefly in the Daisy-order, the anthers are joined together. They are then called *syngenesious*. In form the anthers are either long and parallel, or short and bag-like, but exhibit many, and some very curious shapes. Usually, however, they present an oblong, oval, or heart-shaped form divided into two cells.

Such being the essential organs, we pass to the envelope, protective or attractive. Here a difference occurs between the exogens and endogens, and we will consider the former first. Some exogens are *monochlamydeous*—having a single covering or cloak. Such are the Chenopodiaceæ and

Polygonaceæ—Goose-foots and Docks. In their case the "covering" is either called a *perianth* or "something around the flower," or a calyx, which is a term more properly applied in other cases.

The covering of the flower in the bud is the *calyx*. It is either of one somewhat tubular piece, opening by points at the top, or is in several separate leaves, or *sepals*. The calyx is usually green, rough, strong, perhaps hairy. Its veins are more generally longitudinal, or parallel, than spreading. Its points are not notched. The calyx is either below or above the ovary, the terms *inferior* and *superior*, whether applied to the calyx or the ovary, indicating their relative positions.

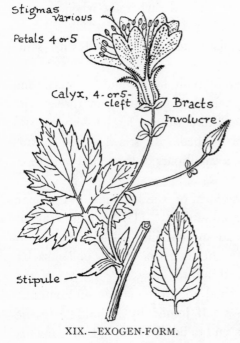

Stigmas various

Petals 4 or 5

Calyx, 4- or 5-cleft

Bracts

Involucre

Stipule —

XIX.—EXOGEN-FORM.

The calyx sometimes falls off—is *deciduous*, or sometimes remains—is *persistent*. In Gooseberry it remains crowning the fruit, to which it is superior. In Tomato it remains about the stalk, and is inferior. In some cases, as in the Rose-order, the fruit develops in connection with it, and the ovary and calyx are there intimately associated.

Behind the calyx, and supporting it, are often a few or many *bracts*, sometimes forming an *involucre* either close beneath the flower, as in the Daisy-order, or some distance below it, as in Anemone. Bracts are often membranous, and somewhat horny, but sometimes are leafy.

The *corolla* is the most showy part of the flower, and occurs between the calyx and the organs in the middle. It is either monopetalous, and perhaps tubular, or consists of separate *petals*. In exogens, the petals and sepals are generally four or five in number, and if the calyx and corolla are each in one piece the segments into which their *limbs* or expanding edges are cut are generally four or five. The petals are usually of a much more fragile texture than the sepals. They arise by narrow claws or stalks, which are sometimes long, from the calyx or from the central body of the flower, below, or above the ovary. They are not unfrequently notched, and are commonly veined with close fine veins expanding to their edges. They are sometimes unequal, and of peculiar irregularity, but when this is the case two are generally different from the other three. In Pelargonium, the "geranium" of the household, the upper two petals are smaller or larger than the lower three. Similar irregularities occur when the corolla is tubular, as is seen in the Scrophulariaceæ, and the division of the five points into two sets of two and three occurs also in the calyxes—as in the Leguminosæ and Labiatæ.

The corolla is the coloured part of the flower, but sometimes colour extends to the calyx. Often the calyx is reddish or purplish, as also are the bracts—in some of the Sages, for instance; the perianth in some Docks is very red; in Tropæolum, the garden nasturtium, the calyx is yellow or orange. But the most notable instance of the calyx being coloured is in the Ranunculaceæ, in which order the sepals very frequently are petaloid, and gaily coloured, so that the sepals are taken for petals—as in Clematis, Anemone and Hellebore.

The petals are sometimes reduced to insignificant or curious forms. In Hellebore they are small green tubular nectaries, in Love-in-a-mist they are small hood-like tubes, in Monkshood they are tiny and quite hidden by the gay sepals. Formerly it was the custom to call all such petals *nectaries*.

We must not overlook the deep spurring which occasionally is seen in both sepals and petals. In sepals—in Columbine, Delphinium, Nasturtium and Canary Creeper. In petals—in Delphinium, Viola, Fumitory, Centranthus, etc.

Such, then, are the flowers. The manner in which they grow on the stem is the *inflorescence*. It varies very considerably. The flower grows, usually, upon a short stalk, or *pedicel*. When there is no pedicel the flower is said to be *sessile*. As a rule the pedicel grows in the axil of (or hollow above) a leaf, or bract, which appears to be only a leaf partaking somewhat of the nature of a sepal. The simplest inflorescence is that in which flowers stand in the axils of leaves alone—as in Pimpernel. But often several flowers grow upon one flower-stalk. This is called a *peduncle*, and the short stalks of the separate flowers are their pedicels. At the base of each pedicel is, as a rule, a bract, from the axil of which it springs. But these bracts are by no means always present. When a peduncle branches and bears several flowers, on their own pedicels, the inflorescence thus formed is called a *panicle*. When a panicle is flattened, so that all the flowers are more or less on one level, it is called a *cyme*—as in Elder and

Laurustinus. When the peduncle rises erect, and sends off flowers on their own pedicels, all the way up, it is called a *raceme*. When the lower flowers have longer pedicels than the upper, and so bring the flowers on to the same level, it is called a *corymb*—as in Candy-tuft. When the pedicels all start from the same point it is called an *umbel*—as in Cowslip. When the flowers up a peduncle have no pedicels, it is called a *spike*. A *catkin*, or *amentum*, is a spike bearing small unisexual flowers—as in Oak, Willow, etc. A *cone* is also a kind of spike—the stiff woody parts of it are carpellary leaves bearing seeds naked upon them—leaves which, if curled up, would enclose the seeds, and be proper carpels. A *head*, or *capitulum*, is such as we see in the Daisy-order, where the tiny flowers are crowded upon a receptacle into a head.

Botanists term some inflorescences indefinite, some definite. In the former the upper or inner part goes on developing and producing buds, those below or at the edge of an umbel, or corymb, coming into flower first. In the latter the stem ends in a terminal flower, and its increase is arrested —as in Anemone and Dandelion. If increase takes place it does so from lateral branches—as in the Lychnis, Chickweeds, etc., where there is a continual terminating and forking of the stem.

It must not be overlooked that in many cases, as in plants of the Wall-flower class, Monkshood, etc., in which lateral racemes and panicles develop later, they are in flower while the central one is in seed. The leaves in exogens have reticulated veins, and their larger veins generally spread out in a palmate fashion. They are often lobed and toothed, and the edges are frequently serrated. The veins run to the tips of the teeth, but not always. Sometimes, as in some of the Scrophulariaceæ, they run to the hollows between the teeth. The veins spread down the leaves in various ways, too many to mention. There are generally some places on the leaf-surface where the veins die away, so far as the easily-seen veins are concerned. These places are continuations of the gulfs between the lobes, and indicate where, in some cases, the leaf has been folded in the bud.

Sometimes the leaves are *pinnate*, or feather-like, bearing small leaflets down either side of the mid-rib. In many cases the blade, or green part of the leaf, creeps as a small border along the edge of the mid-rib, so that the leaflets are connected. Such a leaf is called *pinnatifid*. In some cases, as in the Cabbage-order, the stalk of the leaf is its mid-rib continued down, without any change but that of getting larger. In these instances, some of the blade also fringes it, sometimes irregularly, and sometimes becomes full against the stem, where it expands into auricles, or ears, as is seen in Sowthistle and Leopard's-bane.

Usually the mid-rib is channelled on the upper side, and perhaps even more when the stalk of the leaf can be more properly so denominated. The stalk, or *petiole*, joins to the stem by a thickened end, which usually grips the stem, and leaves a moon-shaped scar when torn off. The leaves fall off at this joint, and though they may be "evergreen," yet that only means that they last a few years instead of for one.

Certainly not to be overlooked are the *stipules* when they are present. They either fringe the lower part of the leaf-stalk, as in Pea and Rose, or stand at its base, as in Saint John's-wort, or are upon the stalk beside the leaf, as in Pelargonium. Sometimes, as in Persicaria, they pass right round the stem. They are usually membranous, or rather horny, more like bracts and sepals than leaves, but they are sometimes leaflike, as in Willow and in some of the Peas.

The stems in exogens vary from the most juicy character to the woodiest and knottiest. Sometimes the bark bears prickles and hairs or bristles. In cross-section, plants of more than a brief life reveal their exogenous character in exhibiting rings of wood, the youngest wood being at the outside.

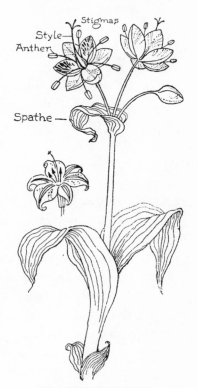

XX.—ENDOGEN-FORM.

The knotty, twisted stems of the exogens we do not see in our *Endogens*. Except in the grasses and some rushes, a certain juiciness is commonly to be seen in these plants.

Their leaves do not fall off as do those of exogens, but wither upon the stem. They are parallel veined, and are unlobed and untoothed. Except in a few cases, they have not even prickly edges. As a rule, too, they have no stalks. But certain exceptions must be noted. The leaves of Arum have branched veins. The leaves of Butomus, Alisma, and their allies have wide leaves with parallel veins, and have stalks. The leaves of Palms are pinnate. In Naiadaceæ and Araceæ there are stipules.

In Grasses the leaf continues down the stalk in a split sheath, and where the blade of the leaf joins its sheath, there is a ligule against the stem, as if to prevent water getting down between the sheath and the stem. In Rushes the sheath is not split. This peculiar transition from tube to flat blade is common in endogens. Compare the form in Iris. In Alstrœmeria, one of the Amaryllidaceæ, the leaves twist round, as if they had been set on upside down and had to twist to get their "upper" surface to the top.

The stems do not branch. Sometimes, as in Hyacinth, they rise up leafless, and bear their flowers in the axils of small bracts, or sometimes they bear leaves all up, the upper ones gradually smaller, and bear the flowers in the uppermost. The stems are often somewhat polygonal.

The flowers differ from exogens in having their parts in threes, and in having a *perianth* of six segments in place of a calyx and corolla. Nevertheless, three of the segments usually act as calyx, and are green to begin with. They usually remain rather smaller than the three inner ones which they have enclosed.

As a rule all the segments have a fairly prominent keel down the back, especially at the bottom—where the inner segments will be seen to make their way to the outside. Sometimes this keel is coloured green, and the outer segments often retain rather a greener hue than the inner ones.

Down the segments inside there is often a line of colour, usually coterminous with three or four lines or veins running down to the point. On either side of these central veins others of very fine character spread out towards the edges. Sometimes the inner and outer segments differ considerably, and two or three are sometimes marked. The outer segments often change from green to the richest of colours. This is particularly to be seen in the Tulips, in which the outer part of each segment is folded in in the bud. But sometimes the outer segments remain green, and perhaps narrow, as in Alisma and Tradescantia. They are then spoken of as sepals, and the gayer inner segments are called petals. In some Gladioli the segments are only four or five.

The ovary is sometimes below the flower—as in Iris, Daffodil, and Snowdrop, or above—as in Lily, Hyacinth, etc. The fruit is three-celled.

Peculiar to the endogens is the *spathe*, which is a membranous covering to the buds—white and handsome in Arum; membranous, brownish and withered in Iris, Daffodil, etc. It is sometimes double, as in Garlick.

The perianth in some plants—as the Lilies and Tulips—drops off when they die, but as a rule both leaves and flowers wither upon the stem.

Classed now with endogens are those plants—such as Black Bryony and Smilax—which Lindley formerly ranked as Dictyogens. Their leaves are net-veined, but usually with a predominance of longitudinal veins. The leaves of exogens sometimes approach those of endogens in their simplicity. Some, as Plantains, have strongly marked longitudinal veins, which would, at first sight, place them among the endogens. Others have grass-like leaves.

In the great class to which the ferns belong are the mosses, lichens, fungi, and sea-weeds.

The leaves of Ferns are fronds. They are either simple and strap-like, as in the common Polypody, or are deeply divided into the pinnate form so immediately connected with them. The main mid-rib of the frond makes a graceful curve, beginning below, broad and shaggy, with brown membranous scales, ragged and woolly. Some measure of these scales are to be found on the back of the mid-rib, all up, and on the backs of the pinnæ as well. The blade of the frond is green, veined with forked veins. In those kinds which have not serrated edges the veins make forks with long prongs, running out to the edge, and more or less at right angles to it. But when the edges are serrated, the branches of the veins run down to the points. The forked character

XXI.—FERN-FORM.

is then largely lost, and in drawing them one has not to put a central line and then branch the lateral veins from it, as one would in an oak-leaf, but must first draw a vein down from the nearest point, then another from the next point to meet it, and so on. The result of this is, that the veins are very gracefully connected with the central ribs of the frond. The young fronds are rolled up in a manner which is called circinate. In those which are pinnate, the lateral divisions are also rolled up towards the mid-rib. The reproduction of the fern is by spores, which are developed in, and fall from, spore-cases upon the backs of the leaves. In some instances they are on the front, and in others—in Osmunda and Ophioglossum—the spore-cases are upon fronds which are given up to that duty.

The draughtsman will notice that the serrations on the edges tend to point to the apex of the frond.   This peculiarity, and the graceful connection of the veins with the mid-ribs, do away with any suggestion of that virile kind of expansion which is the chief characteristic of the exogenous plants.

A note must be made upon the drawing of the turn-over of the leaf. The old draughtsmen followed the rule illustrated in the accompanying diagram.   They broke the nearer outline into two, one over-lapping the other, as it were, leaving a small passage between them as if the leaf were very thick.   This is a very valuable convention, and is practically indispensable to the decorator.

XXII.—HOW TURNED-OVER LEAVES ARE DRAWN.

CHAPTER VI

THE CLASSIFICATION OF PLANTS

IN the early days of plant-study, when our old Herbals were composed, it was customary, if any arrangement at all was adopted, to put together those plants which were similar in use or in habitat. Quite early the writers of those histories of plants began to associate together those plants which bore outwardly the marks of their kinship. The Umbelliferæ were among the earliest to be thus brought together.

But the discovery of sex in plants towards the end of the seventeenth century called attention to the stamens and pistils as of paramount importance. Linnæus closed a series of attempts at classification by the promulgation of that system which goes by his name. The artificial system of Linnæus at least served as a good index to the vegetable kingdom, while the investigations upon which alone a more natural system could be based were being made.

That some system would have ultimately to be adopted which placed plants together according to their kinship rather than according to such a matter as the number and condition of their chief organs, Linnæus himself saw. But it was not till about 1830 that the Natural System was adopted as the basis of classification.

Lindley, in his *Vegetable Kingdom* (London, 1845), put forward a system which, though it is now slightly modified by botanists, is interesting to the artist. He divided the whole plant-world into seven classes, and these again into Alliances. Each alliance included the orders which were closely connected. In this system Lindley began with the simplest plants and ended with the most highly organized. It has become customary to reverse the order, and therefore in the following brief *résumé* of his system it is turned round.

FLOWERING PLANTS.

1. *Exogens.*—Wood of the stem youngest at the circumference, always concentric.

Cotyledons, or seed-leaves, 2.

Veins of the leaves netted.

Flowers with their parts often in 4's or 5's.

Leaves articulated upon the stem, and falling off.

Seeds enclosed in seed-vessels.

Leaves often lobed, and with serrated, or toothed, edges.

The greater part of our common plants belong to this class—Oak, Rose, Daisy, Buttercup, etc., etc.

2. *Gymnogens.*—Wood of the stem youngest at the circumference, in concentric rings.

Cotyledons, 2 or more.

Leaves narrow at the base, jointed to the stem, falling off.

Veins of the leaves parallel, or the leaf too narrow to allow reticulation.

Seeds naked upon a carpellary (or seed-vessel-like) bract.

These are the Pine-trees and their allies. They closely resemble club-mosses and ferns. Lindley said, "So great is the resemblance between Club-mosses and certain Conifers, that I know of no obvious external character, except size, by which they can be distinguished."

3. *Dictyogens.*—Cotyledon, 1.

Leaves net-veined, usually disarticulating from the stem.

Stems approaching the character of exogens.

Flowers often with their parts in threes.

These plants, of which the Common Black Bryony is one, are generally placed amongst endogens, but, as Lindley points out, except for their single cotyledon they might be placed amongst exogens. He claims that they are intermediate between the two classes.

4. *Endogens.*—Wood of the stem youngest in the centre. Wood confused, and not in concentric rings.

Cotyledon, 1.

Leaves with parallel veins, not articulated with the stem, not falling off, but withering upon it.

Flowers with their parts in threes ; or glumaceous (having husks).

Leaves neither lobed, toothed nor serrated.

These are the Lilies and the Grasses. In temperate regions they are herbs only, but in the tropics become great trees. The Arums are exceptions to the rule in regard to the form and venation of the leaves.

5. *Rhizogens.*—Parasitical plants without true leaves, but scales ; without a proper stem, but a fungoid amorphous mass instead. Many botanists objected to these plants being made into a separate class, and Bentley, who followed Lindley's classification to a considerable extent, placed them among the dicotyledons, or exogens. They are mentioned here merely to give Lindley's arrangement undisturbed. The genera are but 21 in number.

FLOWERLESS.

6. *Acrogens.*—Reproduction by spores, in sporagia, or spore-cases, usually on the backs of the leaves in ferns, or upon rudimentary leaves as in Osmunda and Ophioglossum.

Leaves distinguishable from the stem.

Stem increasing at its summit, and, if branching, doing so in a forked manner.

Veins of the leaves forked.

These are the Mosses, Ferns, Horse-tails, etc.

7. *Thallogens.*—Root, stem and leaves confused.

Reproduction by very simple, though by more or less sexual, means.

These are the Sea-weeds, Rose-tangles, Fungi and Lichens.

Such are the classes of plants. It is, however, usual now to name only three : Exogens, Endogens, and Acrogens, using the terms Dicotyledons, Monocotyledons, and Acrogens or Cryptogams.

To the artist the characteristics of these classes are useful. A very slight observation of plants will make clear the distinction between them. Some exogens, as the Plantains, may at first sight appear to be endogens, on account of the strong parallel veins of their leaves. But these instances are few.

The arrangement in alliances which Lindley made appears now to be wholly given up, and it is not, from our point of view, of any service.

The natural system, as we now see it in the botanical works, consists of nearly three hundred Natural Orders, arranged according to their affinity and beginning with the most complex plants and ending with the simplest. The name of a natural order is a compound of the name of a representative plant in it, and of a termination—aceæ—which is apparently the feminine plural of an adjective. So " plantæ ranunculaceæ " are the ranunculus-like plants. This peculiar termination indicates that the word is the name of a natural order, and no confusion seems to have arisen. However, Lindley poured scorn on the number of syllables, and tried to substitute English words, largely of his own construction—as " Rose-worts." Some of these scientific Latin names have been shortened a syllable or two, and Scrophu-larineæ is sometimes used for Scrophulariaceæ.

In common speech, when we wish to designate the relationship of a plant, we speak of it as " a lily " or " a kind of lily," or " one of the lilies." This may, of course, mean that it is one of the genus Lily or Lilium, and not that it is one of the Liliaceæ. Lindley would say it is a Lily-wort—but that word has not got into even the copious dictionaries of to-day.

Each natural order comprises certain Genera, and each Genus certain Species. The species are the individuals. " Varieties " are non-constant variations from the normal character of a species. They can only be main-tained by special cultivation.

Each plant has a generic and a specific name. The first indicates the genus, the second the particular species in that genus. The specific name is properly an adjective, as Ranunculus repens—the creeping Ranunculus. Sometimes it is an ancient name used as a distinction, as Ranunculus Flammula—the Flammula Ranunculus. Sometimes when the bounds of a genus have been shifted, and the genera have been renamed, what was a generic name becomes a specific one—as Coronopus Ruellii, or the Coronopus of Ruellius, became Senebiera Coronopus. When the specific name is a proper noun it has a capital letter.

What peculiarities separate one species from another, and one order from another, is a matter not only far beyond the present writer, but also beyond the present purpose. For all but botanists, the broad distinctions between genera are all that is necessary. Nay, many genera may be run together. But scientific nomenclature is now so common that it cannot be avoided, and as far as possible the figures given in the following pages have been given their scientific names as now determined. Nevertheless, in a great many cases specific and generic distinctions are matters of internal botany, and mean nothing to those who deal with externals. As a rule, however, the main characteristics of an order are distinctly useful to the artist. They help him to render his forms. In the brief descriptions which precede each set of figures, this purpose has been kept in view, and mere botanical descriptions have been avoided as much as possible.

In the identification of the figures recourse has been had to the later Georgian botanists—Smith, Withering, Stokes, Loudon, Martyn, etc.

The arrangement adopted is that given by Bentley in his *Manual of Botany*, which is a useful and a clear statement of both structural and systematic botany. For the English plants the reader is advised to consult Bentham and Hooker's *British Flora*, the woodcut figures in which, by Fitch and Smith, deserve the highest praise.

The only departure from Bentley's arrangement has been in the placing of the Dictyogens, where Lindley put them, between the Exogens and the Endogens. Although on grounds of internal structure they may more correctly be placed where Bentley has them, to an artist they must be placed apart from the parallel-veined plants.

A list follows of the chief Natural Orders according to the arrangement adopted. It will be seen that its subdivisions are based upon the position of the stamens and upon the divided or undivided condition of the corolla. The divided or undivided condition of the calyx is not mentioned, but the progress is, as in the case of the corolla, from divided to undivided. There are, however, fewer plants with a divided calyx, of separate sepals, than there are with a divided corolla of separate petals.

As the arrangement follows such external marks as the position of the stamens and corolla, and other similar matters, the task of finding in which Order a given plant lies, is simplified. But it is not always easy to correctly determine the description of the plant in one's hand. Exceptions have to be guarded against. The chief of these is the absence of petals, which occurs here and there among plants having both sepals and petals.

THE VEGETABLE KINGDOM

FLOWERING PLANTS

I.—EXOGENS OR DICOTYLEDONES. The parts of the Flower usually in 4's or 5's.

(A) DICHLAMYDEÆ. **Calyx and Corolla distinct** (*dis*, two, and *chlamys*, a covering).

(i) THALAMIFLORÆ. **Stamens under the pistil,** upon the *thalamus*. **Petals, separate, inserted into the thalamus.** (*Thalamus*, a couch, a ridge projecting below the ovary, or pistil, bearing the calyx, corolla, and stamens, if all are upon the thalamus, as in this section.)

Note.—Some few Thalamifloræ exceptionally are apetalous, or monopetalous.

RANUNCULACEÆ, Buttercup.	VIOLACEÆ, Violet.	ACERACEÆ, Maple.
MAGNOLIACEÆ, Magnolia.	DROSERACEÆ, Sun-dew.	AURANTIACEÆ, Orange.
BERBERIDACEÆ, Barberry.	TAMARICACEÆ, Tamarisk.	VITACEÆ, Vine.
NYMPHÆACEÆ, Water-lily.	CARYOPHYLLACEÆ, Pink.	RUTACEÆ, Rue.
PAPAVERACEÆ, Poppy.	MALVACEÆ, Mallow.	ZYGOPHYLLACEÆ, Bean-caper.
FUMARIACEÆ, Fumitory.	TILIACEÆ, Lime.	LINACEÆ, Flax.
CRUCIFERÆ, Cabbage.	CAMELLIACEÆ, Camellia, Tea.	OXALIDACEÆ, Wood-sorrel.
CAPPARIDACEÆ, Caper.	HYPERICACEÆ, St. John's Wort.	BALSAMINACEÆ, Balsam.
RESEDACEÆ, Mignonette.	SAPINDACEÆ, Horse-chestnut.	GERANIACEÆ, Geranium.
CISTACEÆ, Rock-rose.	POLYGALACEÆ, Milk-wort.	TROPÆOLACEÆ, Indian cress.

(ii) CALYCIFLORÆ (Calyx-flowers). **Stamens on the Calyx. Petals separate, upon the Calyx.**

Note.—In Cucurbitaceæ, and sometimes in Crassulaceæ and Araliaceæ the Corolla is monopetalous. Some few Calycifloræ are apetalous—as Lady's Mantle.

(*a*) PERIGYNÆ. **Ovary above the Calyx. Flower around the Ovary.**

CELASTRACEÆ, Spindle-tree.	ROSACEÆ, Rose, Apple.	PORTULACACEÆ, Purslane.
STAPHYLEACEÆ, Bladder-nut.	LYTHRACEÆ, Loose-strife.	PASSIFLORACEÆ, Passion-flower.
RHAMNACEÆ, Buckthorn.	SAXIFRAGACEÆ, Saxifrage.	
ANACARDIACEÆ, Sumach.	HYDRANGEACEÆ, Hydrangea.	
LEGUMINOSÆ, Pea, Bean.	CRASSULACEÆ, House-leek.	

Note.—In Rosaceæ and Saxifragaceæ the ovary is sometimes so prominent within the calyx as to appear inferior to it—notably in Roses.

(*b*) EPIGYNÆ. **Ovary beneath the Calyx. Flower above the Ovary.**

CUCURBITACEÆ, Gourd.	PHILADELPHACEÆ, Syringa.	CORNACEÆ, Cornel-tree.
CACTACEÆ, Cactus.	MYRTACEÆ, Myrtle.	UMBELLIFERÆ, Parsley.
GROSSULARIACEÆ, Gooseberry.	ONAGRACEÆ, Fuchsia.	ARALIACEÆ, Ivy.

(iii) COROLLIFLORÆ (Corolla-flowers). **Stamens on the Corolla, except as below. Corolla monopetalous.**

(*a*) **Ovary beneath the Calyx.**

CAPRIFOLIACEÆ, Honeysuckle.	DIPSACACEÆ, Teazel.	LOBELIACEÆ, Lobelia.
RUBIACEÆ, Madder.	COMPOSITÆ, Daisy.	VACCINIACEÆ, Cranberry.
VALERIANACEÆ, Valerian.	CAMPANULACEÆ, Bell-flowers.	

(*b*) **Ovary above the Calyx. Stamens upon the thalamus.**

ERICACEÆ, Heath.	PYROLACEÆ, Winter-green.

(*c*) **Ovary above the Calyx. Stamens on the Corolla.**

EBENACEÆ, Ebony.	POLEMONIACEÆ, Phlox.	BORAGINACEÆ, Borage.
AQUIFOLIACEÆ, Holly.	SOLANACEÆ, Nightshade.	EHRETIACEÆ, Heliotrope.
STYRACACEÆ, Storax.	OLEACEÆ, Olive.	LABIATÆ, Sage.
APOCYNACEÆ, Periwinkle.	JASMINACEÆ, Jasmine.	VERBENACEÆ, Vervain.
GENTIANACEÆ, Gentian.	PRIMULACEÆ, Primrose.	ACANTHACEÆ, Acanthus.
CONVOLVULACEÆ, Convolvulus.	PLUMBAGINACEÆ, Thrift.	SCROPHULARIACEÆ, Figwort.
CUSCUTACEÆ, Dodder.	PLANTAGINACEÆ, Plantain.	

(B) MONOCHLAMYDEÆ. Having a Calyx or Perianth only. Sometimes the calyx or perianth is coloured—as in Mezereon, and looks like a corolla. (*Monos*, one, and *chlamys*, a covering.) Plants which though apetalous yet belong to Orders which normally have a corolla remain in their proper Orders above.

POLYGONACEÆ, Buckwheat.
AMARANTHACEÆ, Amaranth.
CHENOPODIACEÆ, Goosefoot.
SCLERANTHACEÆ, Scleranthus.
THYMELACEÆ, Mezereon.
ELÆAGNACEÆ, Oleaster.
LAURACEÆ, Laurel.
BEGONIACEÆ, Begonia.

ULMACEÆ, Elm.
URTICACEÆ, Nettle.
CANNABINACEÆ, Hemp.
MORACEÆ, Mulberry.
PLATANACEÆ, Plane.
EUPHORBIACEÆ, Spurge.
EMPETRACEÆ, Crowberry.
ARISTOLOCHIACEÆ, Birthwort.

LORANTHACEÆ, Mistletoe.
JUGLANDACEÆ, Walnut.
CUPULIFERÆ, Oak, Hazel, Beech, Chestnut, Hornbeam.
MYRICACEÆ, Gale.
BETULACEÆ, Birch, Alder.
SALICACEÆ, Willow.

[The last five form the Amentaceæ—having catkins.]

(C) GYMNOSPERMIA. "Naked-seeded"—having the ovules not enclosed in ovaries, but bare upon carpellary scales.

CONIFERÆ, Pine. | TAXACEÆ, Yew.

II.—DICTYOGENS. Plants now placed amongst Endogens, but having net-veined leaves.

DIOSCOREACEÆ, Black Bryony. | SMILACEÆ, Smilax.

III.—ENDOGENS OR MONOCOTYLEDONES. The parts of the Flower in 3's.

(A) PETALOIDES. Perianth usually coloured, and arranged in a whorl.

(i) **Ovary inferior. Perianth adherent.**

ORCHIDACEÆ, Orchids.
MARANTACEÆ, Arrow-root.

MUSACEÆ, Banana.
IRIDACEÆ, Iris.

AMARYLLIDACEÆ, Daffodil.
BROMELIACEÆ, Pine-apple.

Note.—The Dioscoreaceæ, among the Dictyogens, above, belong to this section.

(ii) **Ovary superior. Perianth free, and liable to fall off.**

TRILLIACEÆ, Paris.
LILIACEÆ, Lily.
MELANTHACEÆ, Meadow Saffron.

COMMELYNACEÆ, Spider-wort.
JUNCACEÆ, Rush.
ACORACEÆ, Sweet Flag.
PALMACEÆ, Palm.

ALISMACEÆ, Alisma.
BUTOMACEÆ, Flowering Rush.

Note.—The Smilaceæ, among the Dictyogens above, belong to this section.

(iii) **Flowers unisexual. Perianth absent or merely scaly.**

TYPHACEÆ, Bull-rush.
ARACEÆ, Arum.

NAIADACEÆ, Pond-weed.
HYDROCHARIDACEÆ, Frog-bit.

(B) GLUMACEÆ. Perianth replaced by husks, not arranged in a whorl, but overlapping, and alternate.

CYPERACEÆ, Sedges. | GRAMINACEÆ, Grasses.

FLOWERLESS PLANTS .

IV.—CRYPTOGAMIA OR ACOTYLEDONES.

(A) ACROGENS. Rooted plants, increasing at the summit. Stems and leaves distinguishable.

FILICES, Ferns.
EQUISETACEÆ, Horse-tails.

LYCOPODIACEÆ, Club-mosses.
MUSCI, Mosses.

HEPATICACEÆ, Liverwort.

(B) THALLOGENS. Stems and leaves confused in a *thallus*.

FUNGI, Mushroom. | LICHENES, Lichen. | ALGÆ, Sea-weed.

It will not escape observation that the stamens radiate more from a point in the Thalamiflorae than in the other sections, in which their bases are wider and wider apart, because they arise from the calyx, or from the corolla.

The statement that the stamens are on the calyx indicates that the petals also are on the calyx. In the Calyciflorae the calyx bears the corolla and stamens, which therefore surround the pistil, and the plants are called Perigynae. But in some Calyciflorae—the Epigynae—the flower is above the ovary.

It must be noted that a monopetalous corolla does not always appear as such. It is easy to see that the corolla of Foxglove is monopetalous, but there are many monopetalous corollas which have the tube so short, and the lobes so much like separate petals, that to attempt to show the mono-petalous character can only injure the expression of the form. In Veronica this is the case. It must be further noted that in deeply lobed monopetalous corollas there is often a marked prominence where the slits between the lobes end—almost a suggestion of overlapping.

A BLOCK USED BY EGENOLPH OF FRANKFORT

FIGURES OF PLANTS

THE FOLLOWING FIGURES OF PLANTS are, with few exceptions, reproduced from Books of the Sixteenth Century. The references which follow
the names indicate from which books the figures have been taken. A list
of the books is given below. When no reference is given, the figure has
been drawn by the compiler. A few figures are from Lindley's *Vegetable
Kingdom*.

LIST OF THE CHIEF ANCIENT HERBALS

The initials standing before some of the names are those used in the references
beneath the figures.

B. = Otto BRUNFELS. *Herbarum Vivæ Eicones.* Strasburg. 1530–1536.

Bock = Hieronymus BOCK, or TRAGUS. *Kreüter Buch.* Strasburg. 1546.

Charles de L'ÉCLUSE or CLUSIUS (1526–1609), edited various
botanical works, and wrote several.

C. E. = Joachimus CAMERARIUS. *Epitome Matthioli.* Frankfort. 1586.

The figures are some of those prepared by Gesner.

C. S. = Joachimus CAMERARIUS. *Symbolorum et Emblematum Centuriæ
Tres.* Frankfort. 1590.

The figures are etchings, and are in medallion form. In the reproductions here given all the design but the plant is cut away. The
parts excluded are absolutely worthless. Apparently these are
the earliest copperplate figures of plants.

C. v. d. P. = Crispian van de PASSE. *Hortus Floridus.* Arnheim. 1614–1617.

The plates are copperplate engravings, each about 8 × 5 inches.

Col. = Fabius COLUMNA (Fabio Colonna). *Ecphrasis.* Rome. 1604.

The figures are etched. His *Phytobasanos* appeared at Naples in
1592.

49

D. = Jacques D'ALECHAMPS or DALECHAMPIUS. *Historia Generalis Plantarum*. Lyons. 1586–1587. Edited by J. de Moulins (Molinæus). The figures are re-drawn and re-engraved from Dodonæus, Matthiolus, Fuchsius, Clusius, Acosta, etc. Some are original. They number 2686, of which about 400 are used twice over.

Rembert DODOENS or DODONÆUS. *Cruijde Boeck*. Antwerp. 1554.

Rembert DODOENS or DODONÆUS. *Stirpium Historiæ Pemptades Sex*. Antwerp. 1583.

E. = EGENOLPH. Publisher of Frankfort, who issued many books with the same cuts about 1550. They are generally crudely coloured —with hardly any attempt at imitation. Many of them are copied from Fuchsius. One of his publications is the *Kreuterbüch*, of Adam Lonicer, 1557, the blocks of which were later reprinted in *Plantarum, arborum, fruticum et herbarum effigies*. 1562. The cut of a man in a garden, at the head of this list, shows how high a point the work which he had done sometimes reached. He was concerned in the production of many books with important illustrations.

F. = Leonhard FUCHS or FUCHSIUS. *De Historia Stirpium*. Basle. 1542.

F. (1545) = Leonhard FUCHS or FUCHSIUS. *Icones Plantarum*. Basle. 1545. Small versions of the cuts in the *Historia* of 1542.

G. = John GERARDE. *The Herball, or General Historie of Plants*. London. 1597. The blocks are those of Tabernæmontanus.

Conrad GESNER. The figures prepared by him were used by Camerarius and later writers, including Zwinger.

J. = Thomas JOHNSON. *The Herball, etc.*, of Gerarde, enlarged and amended. London. 1633. The blocks are derived from Dodonæus, Lobel and Clusius. Some few are original, and some copied from Fuchsius, and Tabernæmontanus.

L. = Matthias de L'OBEL, LOBEL or LOBELIUS. *Plantarum seu Stirpium Icones*. Antwerp. 1581. These were impressions of the blocks used in his works—*Stirpium adversaria nova*. London, 1570–1571, and *Plantarum seu Stirpium Historia*. Antwerp. 1576.

L. V. K. = John LINDLEY. *The Vegetable Kingdom*. London. 1845. The blocks were done by a process called Glyphography.

M. = Petrus Andreas MATTHIOLUS. *Commentarii in libros sex Pedacii Dioscoridis*. Venice. 1554. A second edition with more figures followed in 1558.

M. (1565) = Petrus Andreas MATTHIOLUS. *Commentarii, etc.* Venice. 1565. The edition with the large figures.

Jacob Theodor of Berg-Zabern (Jacobus Theodorus TABERNÆMONTANUS). *Neuw Kreuterbüch*. Frankfort. 1588–1591. The figures were used in Gerarde, 1597.

William TURNER. *New Herbal*. Cologne. 1551. The figures are Birckmann's copies of the cuts in Fuchsius's *Icones* (1545), with a few others.

Z. = Theodorus ZWINGER. *Theatrum Botanicum*. Basle. 1744. The cuts are some of those prepared by Gesner.

As a rule the figures have been reduced about one-third in scale.

The FIGURES are arranged in their NATURAL ORDERS, which are assemblages of those which have naturally some affinity to one another, although they may differ in size, form, and colour. Usually, however, the plants in a Natural Order have considerable external similarity.

[Here begin the EXOGENS, which are plants with net-veined leaves, etc., and of them, here begin the THALAMIFLORAL EXOGENS, which have flowers with the calyx, corolla and stamens usually all arising separately from the THALAMUS, or summit of the stalk below the ovary.]

RANUNCULACEÆ—CROWFOOTS.

These are acrid or poisonous plants, mostly herbs. Their general character is as follows. The Fruit is a bunch of separate capsules, either one-seeded and not bursting, or many-seeded and bursting down the inner side. The latter kind are comparatively few together, standing back to back in a ring. The former are studded over a receptacle into a somewhat globular or cylindrical head. Styles, one to each ovary, give the fruit a kind of beak. Stigma not clubbed nor divided. Stamens numerous with yellow anthers on white filaments, or with purple or black anthers on white or violet filaments.

Petals and Sepals normally five each, as in Buttercup, but subject to much peculiarity. The petals bear Nectaries, and are sometimes deeply spurred, or are reduced to tubular, or peculiar forms. They are sometimes absent. The sepals, to compensate for the deficiency in the petals, are frequently large and petaloid, and gay in colour.

The petals are roundish, concave, slightly notched at the tip, and are attached to the flower by narrow claws, or stalks, which are, however, of no great length. The sepals are usually deeply concave, often hairy, and less frequently green than of other colours.

The colour of the petals is generally yellow, red, or white, but in a few cases blue, violet, purple, or green. The sepals tend to lose the green colour, and to become yellowish or straw-coloured. When petaloid they exhibit a great range—white, white tinged rose-purple, greenish, yellow, pink and all shades of red, violet, purple and purplish blue. The Leaves are generally palmate, cut into deep lobes, and with large teeth. It is unusual for the edges to be serrated with regular little teeth, as in Caltha (Fig. 30). Often the leaves have three main lobes, or divisions, which are sometimes separate leaflets upon separate stalks, as in Columbine (Fig. 46). In colour the leaves are blackish-green, paler and greener behind. The leaf-stalks are also pale green, channelled above, roundish beneath, and expand below, in a somewhat sheathlike manner, to clasp the stem. There are no stipules.

Perhaps the normal growth of these plants may be said to be, that they have an upright stem with leaves on the lower part, and flowers on the upper. The leaves are larger towards the root, and smaller and simpler in form towards the top—as in Columbine. Above, the stem probably branches into a loose panicle of flowers. In some cases there is no stem—as if the plant had been shut up like a telescope ; the flowers finding their place in the middle, and the leaves all round.

In Anemone there is an involucre of three more or less irregular and membranous leaves placed a little below the flower.

One genus only consists of shrubs—Clematis. It differs from Crowfoots as a whole in having opposite leaves. It climbs by its twining leaf-stalks.

The following genera have petaloid sepals, and therefore have no green calyx behind their flowers :—Clematis, Thalictrum, Anemone, Caltha, Trollius, Helleborus, Aquilegia, Delphinium, Aconitum, Eranthis.

Several plants of this Order are figured in the Appendix :—Anemone, Nos. 18 and 19 ; Clematis, No. 21 ; Green Hellebore, No. 20 ; Liverwort, No. 22 ; Love-in-a-Mist, No. 23 ; Winter Aconite and Baneberry, No. 55.

1.—TRAVELLER'S JOY. LADY'S BOWER. OLD MAN'S BEARD.

CLEMATIS VITALBA. F. 97.

The British Wild Clematis. Its name—Old Man's Beard—alludes to the mass of fruit, each fruit consisting of a number of carpels with long feathery tails. These are ill shown at the top of the figure, but are given in a separate drawing, Fig. 2. The beards are particularly conspicuous in damp weather. Flowers almond-scented. Sepals 4 or 5, white. Petals o. Scale of this figure ¼ full size. July.

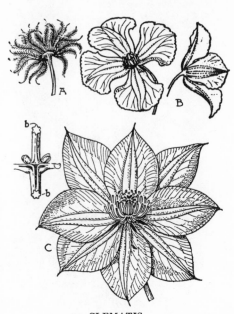

2.—CLEMATIS.
A. Fruit.
B. Clematis Viticella.
Purple. Fl., 3 in. June—Sep.
C. Clematis florida.
White. Fl., 4 in. Ap.—Sep.

3.—MEADOW-RUE.
Thalictrum.
A. Th. flavum. Yellow. 3 ft. July.
B. Th. minus. Purplish. 1 ft. July.
C. Th. alpinus. Pale yellow. 6 in. July.

4.—WOOD ANEMONE.
Anemone nemorosa.
White, pale rose behind. 8 in. Fl., 1 in. April.

5.—AUTUMN ANEMONE.
Anemone japonica.
White. Pale rose-purple. 3 ft. Fl., 2 in. Sep.

53

6.—PURPLE WIND-FLOWER.
G. 302.1.
(See Appendix, Figs. 18, 19.)

7.—MOUNTAIN ANEMONE.
ANEMONE APENNINA. G. 303.5.
Blue. 6 in. Fl., 1½ in. April.

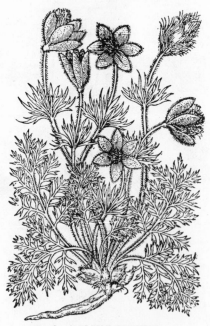

8.—PASQUE-FLOWER.
ANEMONE PULSATILLA. G. 308.1.
Purple. 9 in. Fl., 1½ in. Ap.—May.

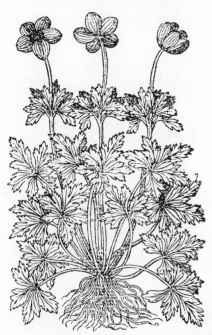

9.—"THE WHITE ANEMONE OF
MATTHIOLUS."
G. 304.8.

10.—PHEASANT'S EYE.
ADONIS AUTUMNALIS.
Red. 18 in. Fl., ¾ in. May—Sep.

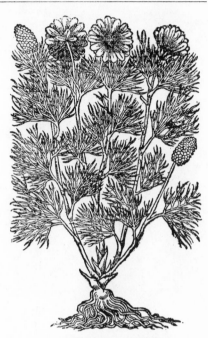

11.—YELLOW ADONIS.
ADONIS VERNALIS. G. 607.1.
Yellow. 18 in. Fl., 1½ in. Mar.—Ap.

12.—CREEPING BUTTERCUP.
RANUNCULUS REPENS. G. 804.1.
Yellow. 8 in. Fl., 1 in. June—Aug.

13.—BACHELOR'S BUTTONS.
RANUNCULUS REPENS. G. 810.1.
Yellow. 18 in. Fl., 1 in. June—Aug.

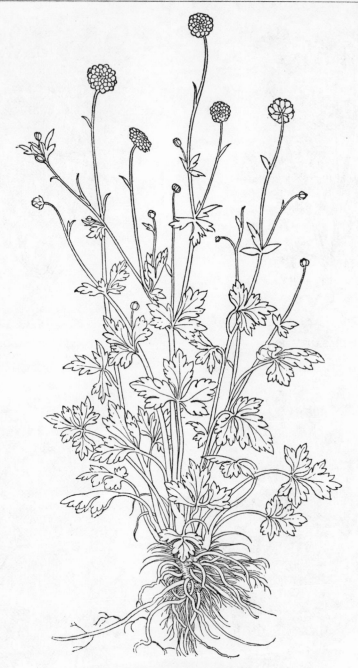

14.—BACHELOR'S BUTTONS.
RANUNCULUS. F. 158.

Bachelor's Buttons are double varieties of R. acris or R. repens. The petals are yellow, but occasionally partly yellow and partly white. There are also exotics with double flowers—R. polyanthemos and R. aconitifolius. The latter is called Fair Maids of France, and has white flowers. Height, 2 ft.

15.—"FROGGE CROWFOOT."
G. 808.9.

16.—"UPRIGHT CROWFOOT."
G. 804.2.

17.—MEADOW CROWFOOT.
RANUNCULUS ACRIS. D. 1032.1.
Yellow. 15 in. Fl., 1 in. June.

18.—WOOD CROWFOOT.
RANUNCULUS AURICOMUS. F. 156.
Yellow. 12 in. Fl., $\frac{3}{4}$ in. Ap.—June.

19.—BUTTERCUP.
RANUNCULUS BULBOSUS.   J. 953.6.
Yellow. 1 ft. Fl., 1 in. Sepals bent back. May.

20.—CORN CROWFOOT.
RANUNCULUS ARVENSIS.   D. 1030.1.
Yellow. 1 ft. Fl., ⅝ in. June.

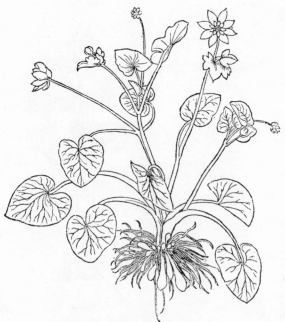

22.—LESSER CELANDINE.
The leaves are splashed with white, and are a paler green behind. The plant often grows in a dense mat. As Gerarde remarks, it wholly vanishes as summer advances. The backs of the petals are partially brown. Sepals 3. Petals 8.

21.—LESSER CELANDINE.
RANUNCULUS FICARIA.   F. 867.
Yellow. 8 in. Fl., 1½ in. Mar.—Ap.

23.—CELERY-LEAFED CROWFOOT.
RANUNCULUS SCELERATUS. M.(1554) 296.1.
Yellow. 18 in. Fl., ½ in. June—Sep.

24.—CELERY-LEAFED CROWFOOT.
RANUNCULUS SCELERATUS. G. 814.4.
Yellow. 18 in. Fl., ½ in. June—Sep.

25.—GREAT SPEARWORT.
RANUNCULUS LINGUA. G. 814.1.
Yellow. 2–3 ft. Fl., 1½ in. July.

26.—LESSER SPEARWORT.
RANUNCULUS FLAMMULA. G. 814.2.
Yellow. 2 ft. Fl., ½ in. June—Sep.

All these are stream-edge or pool-edge plants.

27.—LESSER SPEARWORT.
RANUNCULUS FLAMMULA.   G. 814.3.
Yellow.   2 ft.   Fl., ½ in.   June—Sep.

28.—WATER CROWFOOT.
RANUNCULUS AQUATILIS.   G. 679.

White, yellow at base.   Fl., ⅝ in.   May—Aug.

Water Crowfoot is sometimes regarded as a variety
of the Ivy-leafed.   The latter creeps over mud.   The
former has finely dissected leaves, all beneath the
water.   They are correctly rendered in the lower right-
hand corner of the block, where they are shown branching two ways.   The leaves are sometimes clipped to a
round shape, sometimes are long and trailing.   The
flowers are white, translucent, with a yellow spot at
the base of each petal.

29.—IVY-LEAFED CROWFOOT.
RANUNCULUS HEDERACEUS.   G. 681.1.
White.   Fl., ¾ in.   June—Sep.

30.—MARSH MARIGOLD. KING-CUPS.
CALTHA PALUSTRIS. G. 670.1.
Yellow. 18 in. Fl., 1½ in. May.

31.—LESSER MARSH MARIGOLD.
CALTHA. G. 670.2.
Apparently a variety of the other.

32.—NOBLE LIVERWORT.
HEPATICA TRILOBA. G. 1032.2.
White, Red, Blue, Purple. 6 in. Fl., ⅔ in. Feb.—Ap.

33.—DOUBLE LIVERWORT.
G. 1032.3.
Garden plants, making often a wide mass.
(See Appendix, Fig. 22.)

34.—"DOUBLE RED PÆONY."
PÆONIA OFFICINALIS. J. 981.3.
Crimson. 3 ft. Fl., 6 in. May—June.

35.—DOUBLE WHITE PÆONY.
PÆONIA OFFICINALIS. J. 982.4.
White. 3 ft. Fl., 6 in. May—June.

37.—TREE PÆONY.
PÆONIA MOUTAN.
Red or White. 4 ft. Fl., 8 in. Ap.—June.

CALYX.

The Calyx consists of two deeply concave leaves, A; two smaller ones, B, one of which is somewhat leafy. Behind is a three-lobed bract or leaf.

The SINGLE PÆONY, P. officinalis or P. corallina, is said to grow wild in Steep Holmes, an island in the Severn. Perhaps it no longer is found there. Pæonies are commonly double, sometimes Poppy-like, with a dark spot towards the base of each petal. Many excellent renderings are to be found in Japanese decoration.

36.—"TURKISH PÆONY."
PÆONIA PEREGRINA. J. 982.7.
Crimson. 2 ft. Fl., 5 in. May—June.

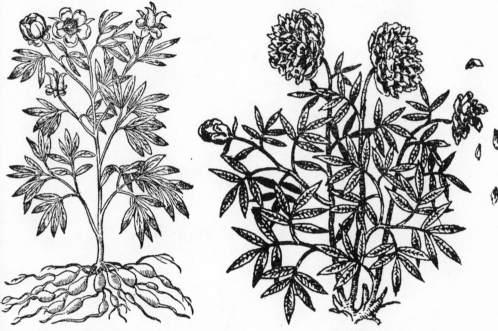

38.—PÆONY.
Pæonia officinalis.　C. E. 658.
Crimson.　2 ft.　Fl., 4 in.　Ap.—May.

39.—PÆONY.　C. S.
This figure is considerably enlarged from an
etching.

40.—"BRANCHED RED ASIAN
CROWFOOT."
Ranunculus.　J. 959.5.
Crimson.　18 in.　Fl., 1 in.　June.

41.—GLOBE FLOWER.
Trollius Europæus.　C. E. 385.
Yellow.　2 ft.　Fl., 1 in.　June.

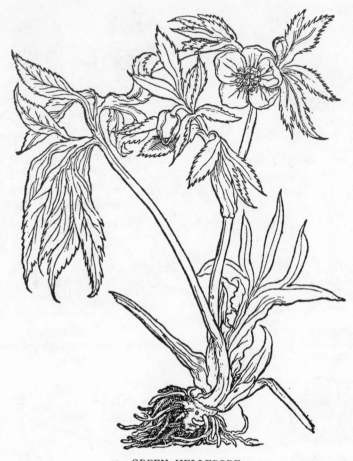

42.—GREEN HELLEBORE.

HELLEBORUS VIRIDIS.　Brunf. i. 30.

This figure is from Brunfel's *Herbarum vivæ eicones*, the first book issued with good figures of plants.　It may well be doubted whether a stronger drawing of a plant has ever been produced.　Compare with it the engraving by Crispian van de Passe, reproduced in the Appendix (Fig. 20).

Of Hellebores, three kinds are common—Christmas Rose (Fig. 45), Green Hellebore, and Setterwort (Helleborus foetidus), which is also called Stinking Hellebore and Bearsfoot.　The last two are British.　In all, the sepals are large and petaloid.　The petals are small, tubular, green, and about twelve in number, and they are perhaps meant to be expressed in the figure above.　The leaf in Hellebores is "pedate"—a peculiar form shown on the next page.　In the Christmas Rose the flowers are on stalks direct from the root, perhaps two on a stalk, with a bract or two, and the leaves also rise from the root and surround the flowers.　In Setterwort there is a stem—the leaves on the lower part, the numerous, pendulous, bell-like lurid flowers on its branches above.　Green Hellebore is between the two. It has slender radical leaves, and its stems branch and bear leaves and flowers as here shown.　The sepals are green, the flower 2 in. across.　The height is 2 ft. Flowers in April or earlier.

43.—GREEN HELLEBORE.

44.—GREEN HELLEBORE. M.(1565)1221.
[This figure has been so identified.]

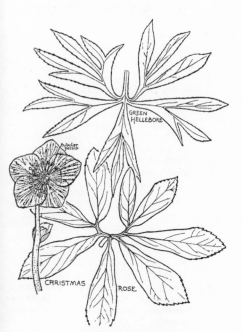

45.—CHRISTMAS ROSE.
HELLEBORUS NIGER.
Leaf 8 in. across. Fl., 2 in. Dec.—Jan.

In CHRISTMAS ROSE the leaves form a low, broad, handsome tuft. Their segments look, from a distance, like leaves of Laurel or Rhododendron. They begin to die down as winter sets in, leaving the middle of the plant exposed, where arise many flower-stalks, about 6 in. high, pale green, tinged with pink, bearing one or two flowers, and having one or two scales or bracts. The sepals are white, tinged behind with "maiden's blush" colour. They remain upon the stalk when the fruit ripens, and become green. The flower forms a hardly regular pentagon.

46.—COLUMBINE.
AQUILEGIA VULGARIS.
Blue, red, white. 2 ft. Fl., 1 in. June.

47.—MONKSHOOD. WOLF'S-BANE.
ACONITUM NAPELLUS.
Blue. 4 ft. Fl., 1½ in. Aug.
B. ACONITUM LYCOCTONUM. Yellow.

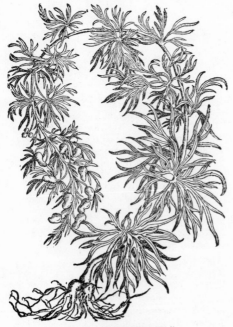

48.—"ACONITUM IX."
M. (1565) 1092.

COLUMBINE.—The name refers to the dove-like " birds " formed by the spurred petals—the sepals representing the wings. The sepals are of the same colour as the petals. The stems are generally reddish or purplish, and so sometimes are the edges of the leaves. There are several garden kinds.

MONKSHOOD.—There are several kinds. The sepals are large, deep blue-purple, or yellow. At the base of the flower are two little pale bracts. The petals are two, small, white, spurred, on long claws, and look like doves drawing a chariot—hence the name "Venus's chariot." Aconite-shaped leaves are often referred to in descriptions of plants with similar leaves. Monkshood is said to be extremely poisonous. Some garden varieties are pale purple.

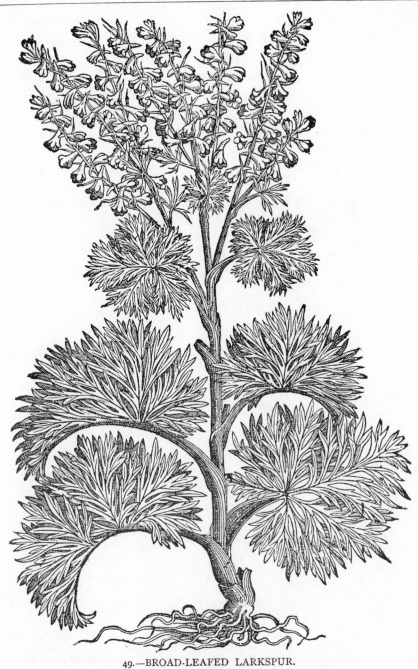

49.—BROAD-LEAFED LARKSPUR.

DELPHINIUM PEREGRINUM (?). M. (1565) 1087.

The same design on a smaller scale is given in Fig. 54, from the first edition of Matthiolus, which was published eleven years before this one. The conclusion is forced upon one that a drawing from nature was the origin of both. Flowers, blue. Height, 1 ft. This is an Italian species, as is also Fig. 48 on the previous page. Both figures are given on account of their finely-rendered leaves.

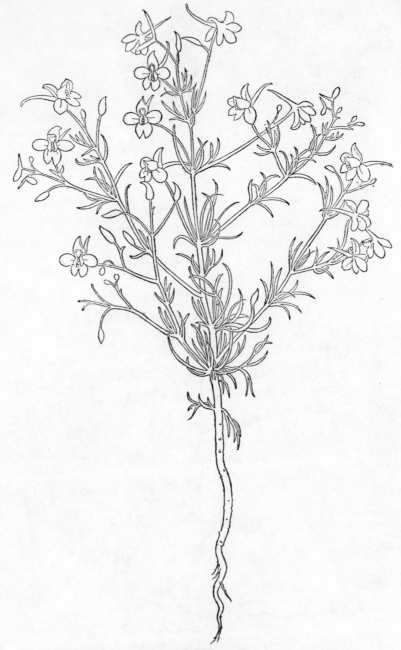

50.—LARKSPUR.
DELPHINIUM AJACIS.　F. 27.

Sepals 5, heart-shaped, the uppermost spurred—blue, pinkish behind, or pink.　Petals 2, of peculiar form, white or pink.　Height 18 in.　Flower, 1 in. or more across.　June—July. The scientific names of this and the Rocket Larkspur, D. consolida, were formerly transposed. The range of colour through flesh-colour, pink, red and blue is very great, especially in the Rocket Larkspur.　In the kind Fig. 53, in which there are 5 petals and the hood-form is lost, the sepals behind them are slightly greener than they, and produce a fine effect of colour.

51.—ROCKET LARKSPUR.
DELPHINIUM CONSOLIDA.　J. 1082.4.
Pink, blue, etc.　3 ft.　Fl., 1¼ in.　July.

52.—LARKSPUR.
DELPHINIUM.　J. 1082.3.

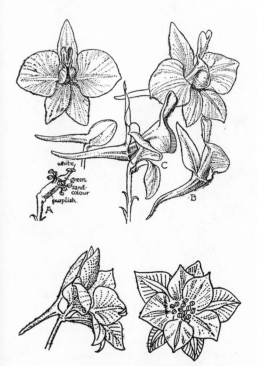

53.—LARKSPUR FLOWERS.

54.—BROAD-LEAFED LARKSPUR.
DELPHINIUM PEREGRINUM.　M. 480 (?).
Blue.　1 ft.　July.

55.—LOVE-IN-A-MIST.

NIGELLA DAMASCENA. F. 504.

Height, 1 ft. Fl., 1½ in. Sepals cold blue or white. Carpels green or bronze-colour. Stamens with purple anthers. South Europe. June—Sep.

For details see Fig. 8.   (See Appendix, Fig. 23.)

56.—LOVE-IN-A-MIST.
NIGELLA DAMASCENA.   D. 813.2.

57.—DWARF NIGELLA.
NIGELLA NANA.   J. 1085.5.
Green-white. Bluish. 6 in. Fl., 1½ in.

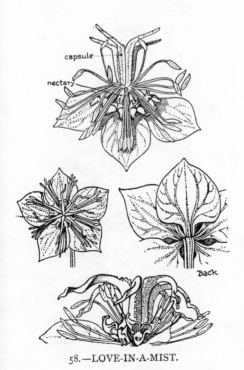

58.—LOVE-IN-A-MIST.

LOVE-IN-A-MIST. — There are several garden kinds, remarkable as much for the horned fruit as for the beautiful flowers, which fading from pale blue to white, with their purple stamens and bronze-green carpels, make a beautiful colour-effect. Other names of the plants are Devil-in-a-Bush, St. Katharine's Wheel or Flower, Bishop's Weed, Fennel Flower.

WINTER ACONITE—Eranthis hyemalis, is a 6-sepaled buttercup-like flower set above a circular involucre. It is one of the earliest garden plants and only 6 in. high.

HERB CHRISTOPHER or BANEBERRY—Actea spicata, is a British plant with black poisonous berries, and small flowers with 4 petals and 4 sepals.

(For these two plants see Appendix, Fig. 55.)

MAGNOLIACEÆ.

Shrubs and trees sometimes of considerable size, bearing large lily-like flowers, usually with 3 sepals and 9 petals, which are generally white or pale greenish-yellow. Natives of North America and China, but cultivated in sheltered gardens. Tulip Tree yields "American white-wood." It has 3 pale-green sepals, and 6 petals, 3 within 3, with an orange mark towards the base.

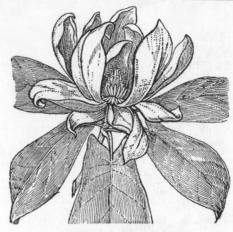

59.—MAGNOLIA AURICULATA.
White. 40 ft. Fl., 4 in. Ap.—May.

60.—MAGNOLIA GRANDIFLORA.
White. 20 ft. Fl., 6 in. June—Oct.

61.—MAGNOLIA CONSPICUA.
White. 30 ft. Fl., 5 in. Feb.—Ap.

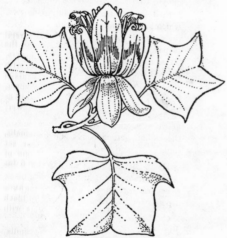

62.—TULIP TREE.
LIRIODENDRON TULIPIFERA.
Pale yellow, orange mark. 60 ft. Fl., 2½ in.
June—July.

63.—PURPLE MAGNOLIA.
MAGNOLIA OBOVATA.
Yellow-white within, purple without. 6 ft.
Fl., 2½ in. Ap.—June.

BERBERIDACEÆ.

Shrubs or herbaceous perennial plants. Flowers
with 3 to 6 sepals, surrounded by scales, with as
many or twice as many petals, and either solitary,
racemose, or panicled, on stalks from the axils
of alternate, compound leaves, usually without
stipules. Chiefly known by the shrubs with rich
yellow or orange flowers, which are common in
gardens. The thorns of Common Barberry are
"nothing more than the hardened veins of leaves,
between which the blade has not developed." Its
berries are used for a preserve, and its bark yields
a yellow dye. Another common shrub is Berberis
aquifolius, with holly-like leaves, tinged with
brown. It bears in August many bluish berries
clustered thickly about its stems. Barren-wort
is a wild British plant. Its red petals bear each
an inflated yellow nectary on the upper side.
Lion's Leaf has its name from the shape of its
leaf resembling a paw.

64.—COMMON BARBERRY.
BERBERIS VULGARIS. Z. 263.
Yellow. 8 ft. Fl., $\frac{3}{8}$ in.

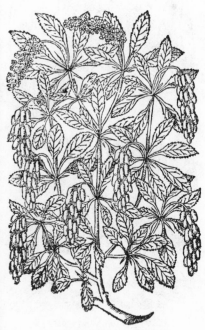

65.—COMMON BARBERRY.
BERBERIS VULGARIS. G. 1144.
Berries, red, $\frac{1}{2}$ in. May—June.

66.—BARREN-WORT.
EPIMEDIUM ALPINUM. L. i. 325.2.
Red, nectary yellow. 10 in. Fl., $\frac{1}{2}$ in.
April.

67.—LION'S LEAF [BERBERIDACEÆ].
LEONTICE LEONTOPETALON. M. 386.2.
Yellow. 1 ft. Levant. Ap.—May.

NYMPHÆACEÆ—WATER-LILIES.

Herbs growing in quiet waters—the leaves and flowers lying on the surface. The fruit consists of many cells arranged in radiating fashion within a turnip-like receptacle, upon which are the sepals, usually 5, and the numerous petals and stamens, in rings one above another. In Nymphæa the receptacle is clothed in this manner to the top; in Nuphar the upper part is bare. The receptacle is crowned by a radiating disc-like stigma, similar to that of the poppy. The exotics include the Egyptian Lotus and the Victoria regia; and have pink, red, blue or white flowers. Sometimes the petals are tinged from white below to some colour at the tips. The edge of the leaves of Victoria regia are turned up. The sepals of Lotus, and the lower petals, are striped yellow and green. Its flower opens through the night.

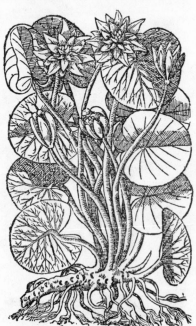

68.—WHITE WATER-LILY.
NYMPHÆA ALBA. G. 672.1.
White. Fl., 4 in. June—July.
(See Fig. 70.) The flower rests upon the water.

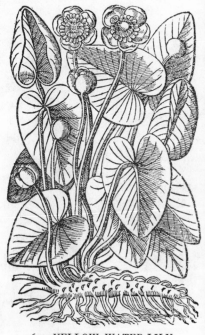

69.—YELLOW WATER-LILY.
NUPHAR LUTEUM. G. 672.2.
Yellow. Fl., 2½ in. July.
The flower rises a little above the water.

PAPAVERACEÆ—POPPIES.

Herbs with a milky narcotic juice, sometimes coloured. Leaves alternate, usually of simple outline, but deeply lobed in a more or less irregular and sometimes pinnatifid manner, and without stipules. Flowers upon long stalks, the buds drooping. Sepals, 2 or 3, falling as the bud opens, and before it opens forming an egg-shape—the petals peeping out between them. The full-blown flower has, therefore, no calyx. Petals, 4 or 6, crumpled in the bud, and in the commonest kinds of a silky fragile texture. Colour : red, white, yellow, orange, violet or deep purple, not blue—very often with a dark spot towards the base. When the petals are large they are distinctly 2 within 2, or 3 within 3.

The fruit is either a capsule or a long pod, very like those of the Cruciferæ. The capsule of the Opium Poppy is well known in its dry state as the Poppy-Head. The capsules are surmounted by a radiating mass of stigmas.

The stem is erect. Often the plants are of a blue-grey, or glaucous, hue.

The rich poppies of the garden are varieties of the Opium Poppy, the Corn Poppy, and of the Oriental Poppy—Papaver orientalis—which has large brilliant petals with a black spot at the base. It is one of the few perennials. Another common exotic is the Norway or Iceland Poppy—Papaver nudicaule—from its long, leafless flower-stalks. It is white, yellow, or orange. The Prickly Poppy—Argemone mexicana—has 6 yellow petals. Its leaves are prickly and have milky veins and edges, like those of Milk Thistle. Another common garden plant is Eschscholtzia, with four yellow petals, and very finely divided leaves.

The wild common Celandine has its flowers in umbels.

Poppies are further figured in the Appendix (Figs. 24 and 53).

70.—EGYPTIAN LOTUS [Nymphæaceæ].
Nymphæa Lotus.
White, tinged pink. June—Sep.
A. Flower of Nymphæa alba.

71.—PRICKLY POPPY.
Argemone mexicana.
Yellow.　2 ft.　Fl., 3 in.　July—Aug.

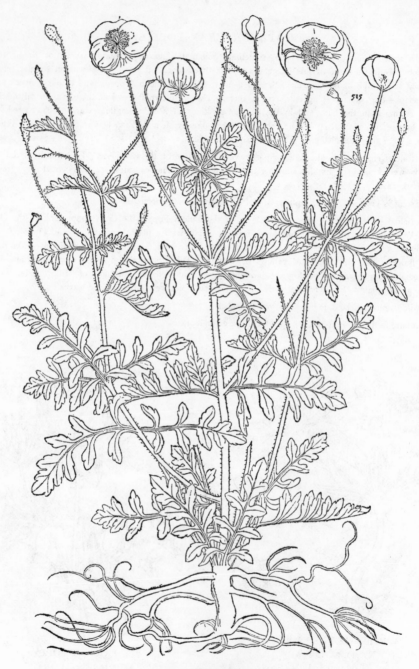

515

72.—CORN POPPY.
Papaver Rhœas. F. 515.
Petals scarlet, with a dark spot towards the base.　2 ft.　Fl., 2 in.　June—Aug.

73.—CORN POPPY.
PAPAVER RHŒAS.   M. (1565) 1057.

[Compare the next figure with this.   This figure was published eleven years after it, and it is a nice question to determine whether the larger cut was an amplification of the smaller, or whether a drawing was the original of both.   Against that theory is the fact that the figures are reversed, unless we can believe that the present one was traced by some process which involved reversal upon the block, whereas the other was sketched upon the block. There is indeed some temptation to think that these large cuts of Matthiolus' were done of the same size as the original drawings.   This reproduction is about 2 in. less in height than the woodcut itself.]

77

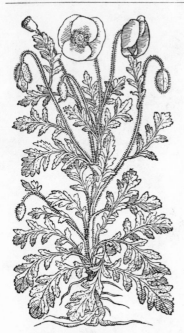

74.--CORN POPPY.
PAPAVER RHŒAS.   M. 469.1.

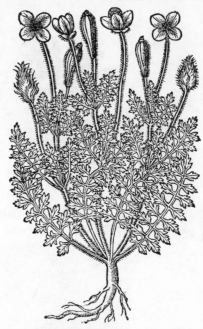

75.—LONG PRICKLY-HEADED POPPY.
PAPAVER ARGEMONE.   G. 300.2.
Scarlet.   18 in.   Fl., 1¼ in.   June.

76.—OPIUM POPPY.
PAPAVER SOMNIFERUM.   M. 469.2.
White, purple spot.   4 ft.   Fl., 3½ in.   July.

77.—DOUBLE POPPIES.
PAPAVER SOMNIFERUM.   D. 1709.2.
(See also Appendix, Fig. 24.)

78.—YELLOW HORNED POPPY.
GLAUCIUM LUTEUM. F. 520.
Pale yellow; leaves and stems blue-grey. 2 ft. Fl., 2 in. Sea-shores. July—Aug.

79.—RED HORNED POPPY.
Glaucium phœniceum. G. 294.2.
Scarlet, black spot. 2 ft. Fl., 1½ in. June—July.

80.—COMMON CELANDINE.
Chelidonium majus. F. 866.
Yellow. 2 ft. Fl., ¾ in. May—June.

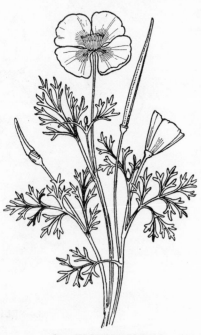

81.—COMMON CELANDINE.
Chelidonium majus.
Yellow. 2 ft. Fl., ¾ in. May—June.

82.—ESCHSCHOLTZIA.
Eschscholtzia crocea.
Yellow, orange spot. 9 in. Fl., 2 in. July—Aug.

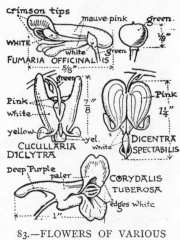

83.—FLOWERS OF VARIOUS SPECIES.

FUMARIACEÆ—FUMITORY.

COMMON FUMITORY (Fr. Fumiterre—L. Fumus terræ, smoke of the earth) is a fragile little plant holding itself in a little roundish bush, or climbing; of a glaucous hue, and by its delicacy of parts not inaptly likened to a little cloud of smoke. The name is said, however, to refer to the odour, or, again, to some magical properties in the smoke of the plant. The leaves are mostly alternate, with the clusters of flowers opposite to them. The segments of the leaves are pointed ovals, the stalks angular in section. The flower has a calyx of 2 small deciduous sepals, and a corolla of 4 petals, 2 within 2. The 2 outer petals are lipped somewhat after the manner of snap-dragon, and the upper one is pouched behind. The flower is rose-coloured, deep purple-red at the lips, with a green keel to the upper and under petals.

RAMPING FUMITORY is larger in every part. Its flowers are paler, though the lips are deep purple. Sometimes white flowers and purplish ones are seen on the same plant. It climbs over other plants by means of its twining leaf-stalks. Plants of the genus Fumaria have for fruit round, or slightly-pointed, nuts. In the genus Corydalis the fruit is a compressed pod.

Formerly the terms appropriate to papilionaceous flowers—banner, wings, keel—were applied to the petals.

There are several well-known garden and greenhouse exotics, notably Bleeding Heart (Dicentra (Dielytra) spectabilis).

In this Order the 6 stamens are in two "brotherhoods," or sets of 3 each. The 3 are connected together by a translucent membrane and produce that peculiar darkened effect at the tips of the petals.

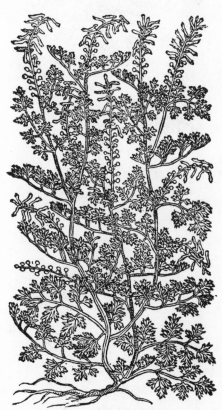

84.—COMMON FUMITORY.
FUMARIA OFFICINALIS.
Pink. 1 ft. Fl., $\frac{1}{4}$ in. May—Aug.

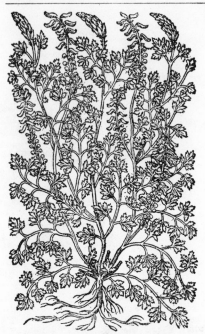

85.—RAMPING FUMITORY.
FUMARIA CAPREOTATA. G. 927.2.
Pale purple. 3 ft. Fl., ¾ in. May—Aug.

86.—YELLOW CORYDAL.
CORYDALIS LUTEA. G. 928.4.
Yellow. 6 in. Fl., ⅝ in. May—June.

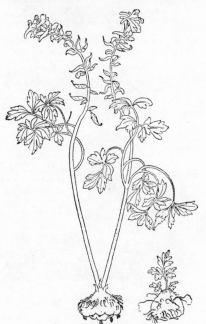

87.—HOLLOW ROOT.
CORYDALIS TUBEROSA. F. 91.
Purple, white. 6 in. Feb.—April.

88.—HOLLOW ROOT.
CORYDALIS TUBEROSA. G. 931.3.

CRUCIFERÆ—CABBAGE, WALL-FLOWER, ETC.

This Order is distinguished by its 4 petals, forming a cross, and by its tetradynamous stamens—6 stamens, of which 4 are long and 2 short. The Order is sometimes named after the genus Brassica, the Cabbage, and one or other species of that genus will be found to fairly exhibit the characteristics of the Order; in which considerable uniformity reigns. A cruciferous plant sends down a tap-root, which is, as it were, a prolongation of the stem downwards. The root sometimes takes the swollen form so well known in the Radish and Turnip. In many plants there is a tuft of radical leaves forming the base of the plant upon the ground. From the middle of the tuft arises usually a single stem bearing leaves of a different character from those at the root. These stem-leaves are often sagittate, having a base with two points, one on either side of the stem. The stem terminates in a corymb, raceme, or panicle, of flowers. The garden Candy-tuft presents a good example of a corymb—the lower flowers on the stalk being on longer pedicels, so that all the flowers are more or less on the same level. In many plants of this Order the flower-stalk continues to lengthen, so that the flat corymb, which it first presents, gradually becomes a tall raceme. This is often very evident in Shepherd's-Purse, the stalk of which will become of considerable length. In many cases but few flowers on the raceme are in bloom together—those below having gone to seed; but sometimes many flowers are perfect at the same time, and then the raceme may be more properly so called, as in Honesty, Rocket, and Turnip. The central flower-stalk is the peduncle, the little flower-stalks are the pedicels. These are arranged spirally one beyond another, not opposite. The peduncle is somewhat ribbed, as if a rib supported a pedicel. There is no little bract where the pedicel branches away, which it does at an upward angle. The flowers do not droop, nor do the fruits which succeed them, except in Woad.

The flower has a calyx of 4 sepals. Crucifers have been divided into those which have a gaping calyx and those which have a calyx with its mouth close. Sometimes the sepals are very equal and regular, but sometimes two overlap and are rather lower than the other two, and are pouched at the base. This is the case in Wall-flower. The petals are sometimes broad, as in Wall-flower and Lady Smock, slightly notched, and with an outline which is of a beautiful but not quite simple shape. Or the petals are narrow though large, as in Honesty, or narrow and small. In colour the petals are generally purple, yellow or white. Rose-colour, flesh, and red also occur. The sepals sometimes approach the petals in colour, as in Wall-flower.

The fruit is also remarkable, and the Order has been divided into two great divisions—the Siliquosæ and the Siliculosæ. A silique is a long pod, longer than broad; a silicule, or pouch, is as broad as long. In both cases there are two valves, one on either side of a central frame. The valves are very deeply pouched in the silicules. They always open from below. A few plants are exceptional, and have a fruit of a different form: Radish has a lomentum, which is a kind of pod, dividing transversely into separate cells. Kale (*Crambe*) has a fruit of two indehiscent cells, the upper of which becomes globose, the lower is abortive and resembles a pedicel. Sea-rocket (*Cakile*) has a similar fruit. Wart-cress has a two-lobed undivided fruit.

Sometimes the stem of the plant branches, sometimes the flowering-branch is panicled. But generally other racemes of flowers arise from the axils of the upper leaves on the main stalk. Such auxiliary flower-branches, although lower on the plant, flower later—are, as it were, a surplus growth. But in Rocket the central raceme and several lateral ones are in bloom together, making a brave show.

The leaves are alternate without stipules, simple in main outline, generally with some slight teeth, on the edge, but not serrated, though usually cut into a waved or jagged edge. Sometimes the leaves have a pinnate character. Usually there is a large lobe at the end, and the blade is continued down either side of the stalk in unequal, irregular, and not opposite lobes.

The middle of the flower reveals the stamens with their yellowish anthers. The style is simple, the stigma simple, but sometimes slightly lobed, and is seen crowning the fruit. Sometimes the style is very short or absent.

In the garden Gilliflower the petals droop. In Rocket they are twisted a little so that the cross is not one formed by right-angles. In Candy-tuft the outer flowers in the corymb have their outer petals larger. The early growth of Rocket is remarkably different from its mature form. Its leaves are very close one above another, and make a rich star form in plan—as if the leaves were in 3's. In the middle is a dense mass of green buds, smallest towards the middle, which, surrounded as it is by small leaves, looks like a green "daisy."

Certain plants of this Order are given in Figs. 3 to 7 and Fig. 53 of the Appendix.

89.—STOCK.

MATTHIOLA INCANA. F. 315.

Purple, Flesh-colour, or White. 1 to 2 ft. Fl., ¾ in. May—July.

Double varieties are very common in gardens. GREAT SEA-STOCK, Matthiola
sinuata, is a somewhat similar plant, 2 ft. high, with downy, wavy leaves, growing
on sandy sea-shores, and flowering from May to August.

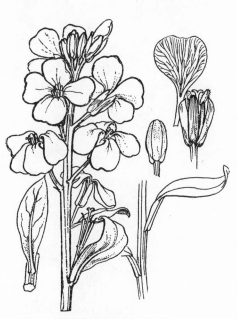

90.—GARDEN WALL-FLOWER.
The whole bud purplish.   Petals, orange, mottled with brown, or rich brown-red.

91.—WALL-FLOWER.
CHEIRANTHUS CHEIRI.   M. 407.2.
Orange-yellow.   18 in.   Fl., 1 in.   Ap.

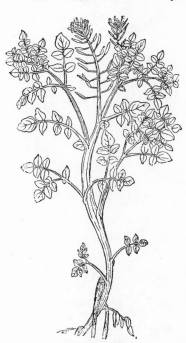

92.—ST. BARBARA'S HERB.
BARBAREA OFFICINALIS.   F. 746.
Yellow. 3 ft. Fl., ¼ in. May—Sep.

93.—WATER-CRESS.
NASTURTIUM OFFICINALE.   F. 723.
White.   9 in.   Fl., ¼ in.   July.

94.—OUR LADY'S SMOCK.
CARDAMINE PRATENSIS. F. 1545.
Lilac. 1 ft. Fl., ⅝ in   May.

95.—TREFOIL LADY SMOCK.
CARDAMINE TRIFOLIA. L. 211.1.
White. 18 in. Switzerland. Mar.—Apr.

96.—SEVEN-LEAVED DENTARIA.
DENTARIA PINNATA. G. 834.3.
Pale purple. 1 ft. Switzerland. May—June.

97.—DAME'S VIOLET. ROCKET.
HESPERIS MATRONALIS. G. 376.2.
White, Rose. 3 ft. Fl., ⅝ in. May—June.

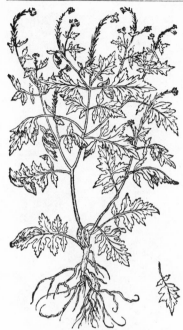

98.—HEDGE MUSTARD.
SISYMBRIUM OFFICINALE. D. 1335.2.
Yellow. 2 ft. Fl., ¼ in. June—Aug.

99.—LEAF OF HEDGE MUSTARD.

100.—FLIXWEED.
SISYMBRIUM SOPHIA. D. 1146.2.
Yellow. 2 ft. Fl., ¼ in. July—Aug.

101.—LONDON ROCKET?
SISYMBRIUM IRIO. G. 189.2.
Yellow. 2 ft. Fl., ¼ in. July—Aug.

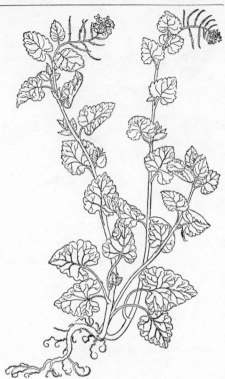

102.—HARE'S-EAR TREACLE MUSTARD.
ERYSIMUM ORIENTALE. L. i. 396.2.
White. 2 ft. Fl., ¼ in. June.

103.—SAUCE ALONE.
ALLIARIA OFFICINALIS. F. 104.
White. 3 ft. Fl., ¼ in. May.

ARABIS.—A pubescent species is very common in gardens, growing in close patches or mats. The leaves are rather thick and very downy, and consequently greyish in colour. The flowers are fairly large. The annexed cut, Fig. 104, does not show the flowers sufficiently all round the stalk, as in all these plants they should be.

104.—ROCK CRESS.
ARABIS ALPINA. J. 275.
White. 6 in. Fl., ½ in. May.

105.—SEA CABBAGE.
Brassica oleracea.   D. 524.4.
Lemon.   2 ft.   Fl., ⅝ in.   June.

106.—A COLE-WORT.
G. 248.15.

107.—A COLE-WORT.
G. 244.5.

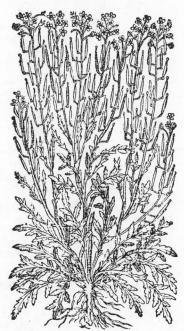

108.—SAND ROCKET.
Brassica muralis.   D. 650.1.
Yellow.   1 ft.   Fl., ⅛ in.   Aug.

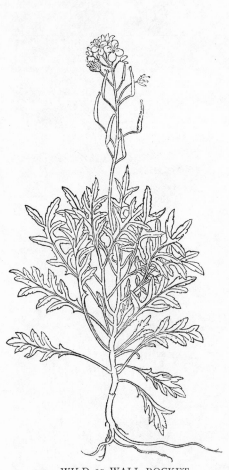

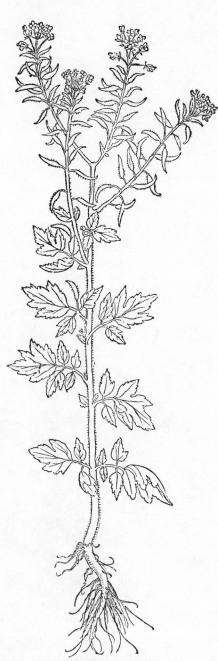

109.—WILD or WALL ROCKET.

BRASSICA TENUIFOLIA. F.

Yellow. 18 in. Fl., ½ in. July—Sep.

SINAPIS.—Several species of Brassica formerly constituted a genus Sinapis—the mustards. Of them the following are here illustrated. Sand Rocket, Fig. 108 ; Wild Rocket, Fig. 109 ; White Mustard, Fig. 110 ; Black Mustard, Fig. 111 ; and Charlock, Fig. 113, which last was called Sinapis arvensis.

110.—WHITE MUSTARD.

BRASSICA ALBA. F. 538.

Yellow. 1 ft. Fl., ½ in. July.

III.—COMMON or BLACK MUSTARD.
BRASSICA NIGRA. F. 539.
Yellow. 3 ft. Fl., ½ in. June.

112.—TURNIP.
Napus sativus. Brassica Rapa. F. 212.
Bright yellow. 2 ft. Fl., ½ in. Ap.—May.

113.—CHARLOCK, OR WILD MUSTARD.
BRASSICA SINAPIS. F. 257.
Yellow. 2 ft. Fl., ⅔ in. May—June.

BRASSICA.—This genus includes the Cabbage, Cauliflower, Broccoli, Borecole, or Greens, Cole-Rape (all which are supposed to have been cultivated from B. oleracea), Colza, B. Napus, the Turnip, B. Rapa, and the Swedish Turnip or Swede, which is from B. campestris, or is a hybrid.

114.—RAPE, OR WILD NAVEW.
BRASSICA NAPUS. F. 177.
Yellow. 2 ft. Fl., ⅔ in. May—June.

115.—FIELD CABBAGE.
BRASSICA CAMPESTRIS. G. 179.1.
Yellow. 2 ft. Fl., ½ in. June—Aug.

116.—SAVOY CABBAGE.
BRASSICA SABAUDA. L. i. 244.1.

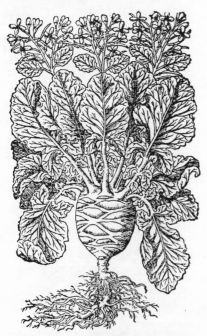

117.—COLE-RAPE, OR KOHLRABI.
BRASSICA OLERACEA GONGYLOIDES. G. 250.1.
A continental cultivated species.

118.—SEA-KALE.
CRAMBE MARITIMA. J. 315.15.
White. 2 ft. Fl., ⅝ in. June.

119.—HORSE-RADISH.
Cochlearia Armoracia. D. 636.2.

120.—HORSE-RADISH.
Cochlearia Armoracia. G. 187.1.
White. 3 ft. Fl., $\frac{2}{3}$ in. June.

121.—SCURVY GRASS.
Cochlearia officinalis.
White. 9 in. Fl., $\frac{2}{3}$ in. May.

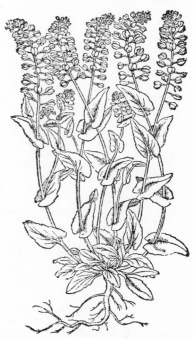

122.—PERFOLIATE PENNY CRESS.
Thlaspi perfoliatum. G. 208.1.
White. 6 in. Fl., small. May.

123.—DITTANDER.　BROAD-LEAFED PEPPERWORT.
Lepidium latifolium.　F. 484.
White.　2 ft.　Fl., small.　July.

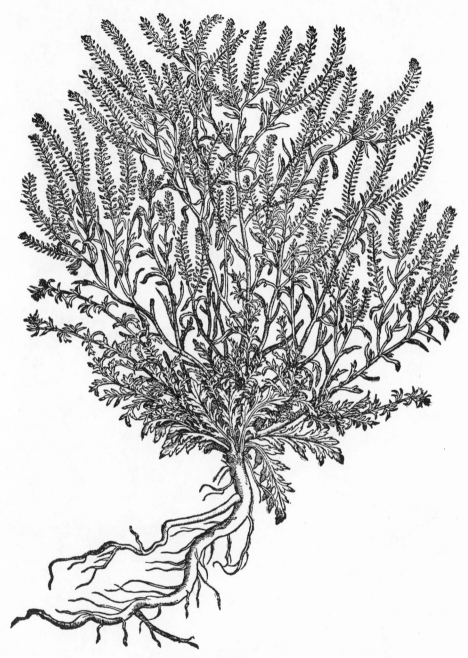

124.—NARROW-LEAFED PEPPERWORT.
LEPIDIUM RUDERALE. M. (1565) 293.
1 ft. Flowers without petals. June. See Fig. 125.

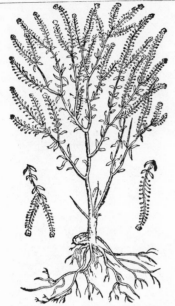

125.—NARROW-LEAFED PEPPERWORT.
D. 662.3.

126.—BOWYER'S MUSTARD.
G. 204.4.

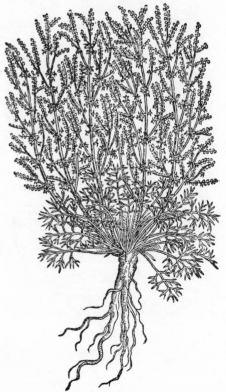

127.—SHEPHERD'S PURSE.
CAPSELLA BURSA—PASTORIS. F. 611.
White. 18 in.

128.—LESSER SWINE'S CRESS.
SENEBIERA DIDYMA. M. 401.1.

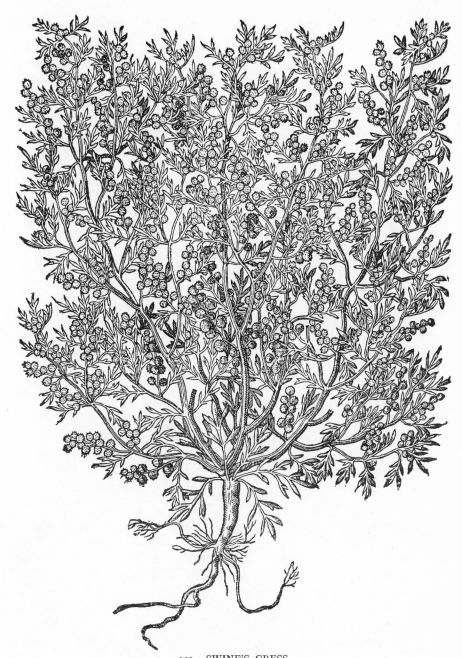

129.—SWINE'S CRESS.

SENEBIERA CORONOPUS. M. (1565) 851.

This and the Lesser Swine's Cress are plants of only a few inches in height. They flower all the summer, and have white flowers.

130.—WOAD.
ISATIS TINCTORIA. F. 331.
4 ft. Fl., yellow. Cornfields. May—July.
Formerly much used for producing a dark-blue dye.

CAPPARIDACEÆ—CAPERS.

An Order of chiefly tropical plants without true stipules, but with sometimes spines in their place. They are similar in many ways to the Cruciferæ, but usually with very many stamens. Sepals and petals, 4. The flower-buds of Common Caper are used for pickle. Cratæva is a hot-house shrub.

CAPPARIS SPINOSA.—An under-shrub. 3 ft. White. Stamens with purple filaments, and yellow anthers. Ovary on a long style-like gynophore. S. Europe. Green-houses. May—June. (For a drawing of the flower see Appendix, Fig. 53.) It grows horizontally from crevices. Stipules spinous.

131.—COMMON CAPER.
CAPPARIS SPINOSA. G. 748.2.

RESEDACEÆ—MIGNONETTE.

Soft herbaceous plants or small shrubs with alternate entire, or pinnately divided leaves, with minute gland-like stipules. The flowers are green —the colour effect modified by the brown anthers of the stamens. The sepals and petals are of rather confused form. The petals bear small finger-like extensions, which in Sweet Mignonette (see Appendix, Fig. 55) form gracefully shaped projections resembling the Honeysuckle ornament of the Greeks.

Sweet Mignonette (Reseda odorata) is a common garden flower which gardeners manage to get in bloom every month of the year.

Reseda lutea and Reseda Luteola are British species.

132.—DYER'S ROCKET. WELD.
RESEDA LUTEOLA. G. 398.1.
Green-yellow. 3 ft. July—Aug.

CISTACEÆ—ROCK-ROSES.

The British Rock-Roses are low, shrubby herbs, with very small, narrow leaves and narrow leaf-like stipules, and with flowers with fine yellow or white petals, not unlike Buttercups at first sight. Cultivated in gardens, are exotic shrubs from 3 to 6 feet high. These bear such a profusion of flowers that though they fall very soon, the shrubs present a handsome show for several weeks.

The sepals are 5, of which 3 are large, and are striated green and white, and 2 are small, behind. The petals are 5, regularly overlapping, somewhat crumpled, and soon falling. Stamens many, style 1. The leaves are simple in outline, often with three strong longitudinal veins. They are opposite, but not always, and vary from narrow to roundish. Some have stipules, and our common Rock-Rose may readily be known by them. There are four British species—one, the White Sun-Cistus, Helianthemum polifolium, has white flowers.

COMMON ROCK-ROSE.

133.—COMMON ROCK-ROSE.
HELIANTHEMUM VULGARE. L. ii. 116.1.
Yellow. 6 in. Fl., 1 in. July—Aug.

134.—COMMON ROCK-ROSE.
HELIANTHEMUM VULGARE. G. 1100.4.

135.—THE "MALE" CISTUS.
M. (1565) 176.

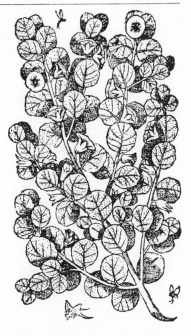

136.—ROUND-LEAFED CISTUS.
D. 223. 1.

137.—MONTPELIER ROCK-ROSE.
CISTUS MONSPELIENSIS.　M. 106.

138.—MYRTLE-LEAFED ROCK-ROSE.
G. 1105. 10.

The height of these is probably 2 feet.　The flowers white or purple.

139.—GUM CISTUS.

Cistus ladaniferus.　M. (1565) 180.

Flowers white, with a purple spot towards the base, and a yellow eye and stamens.　6 ft.　This and
other species yield the fragrant gum—Ladanum.　South of France.　Gardens.　June—July.

VIOLACEÆ—PANSIES AND VIOLETS.

The Pansy is one of the species of Violet. It differs from the others in being parti-coloured. The common colour of Violets is purple, purplish-blue, or violet. But a tinge of yellow invades the spur, and white invades the spur in Dog Violet sometimes, while in Pansy the lowest, or the three lowest, petals are yellow. Also, the Yellow Pansy (Viola lutea) has all the petals yellow, and the many varieties in the garden are well known. The petals are 5, the lowest spurred. The calyx is of 5 sepals projecting backwards as well as forwards, forming a ring of blunt leaves around the stalk, which curves gracefully over the spur. The fruit opens by three valves, the lips of which, on contracting, shoot the smooth seeds a considerable distance. The leaves are heart-shaped, with stipules, which in Pansy are large. Few plants have received more names than the Pansy (*pensée*—thought). A few of them are—Herb Trinity, Three-faces-under-a-hood, Love-in-Idleness, Butterfly Violet, Step-mother. The petals of Sweet Violet are sometimes long and rather strap-like. The plant does not send up a stem; it creeps by runners. Dog Violet makes a stem, which sometimes extends to a considerable length. There are several British species of Violet. Some of the flowers have no petals. They are more generally fertile than the others.

140.—" YELLOW VIOLETS."
G. 700.5.

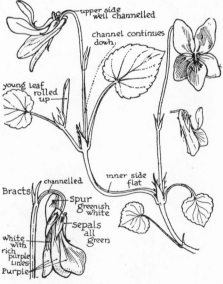

141.—DOG VIOLET.
VIOLA CANINA.
Blue-purple. 6–24 in. Fl., 1 in. Ap.—Aug.

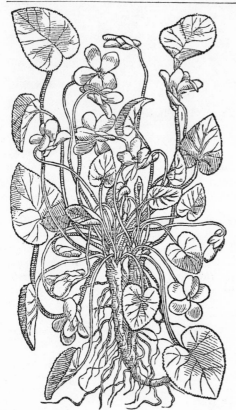

142.—SWEET VIOLET.
Viola odorata. M. 511.

144.—PANSY.
L. i. 611.1.

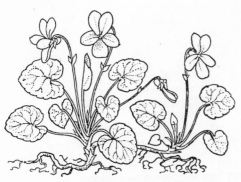

143.—SWEET VIOLET.
Viola odorata.
Purple, white. 4 in. Fl., ⅜ in. Mar.—Ap.

145.—PANSY. HEARTSEASE.
Viola tricolor. G. 704.3.
For colour, see Fig. 147. 9 in. Fl., ⅝ in.

146.—"STONIE HEART'S-EASE."
G. 704.4.

147.—GARDEN PANSY, VIOLA LUTEA,
AND WILD PANSY.

DROSERACEÆ—SUNDEW.

Marsh plants with a scorpioid inflorescence, and chiefly remarkable for the irritability of their leaves. Both the species here figured are fly-catchers, the leaves folding up when touched and imprisoning insects till they are absorbed.

148.—SUNDEW.
DROSERA ROTUNDIFOLIA. Lind. 433.
White, hairs red. 4 in. Bogs.
July. Br.

149.—VENUS FLY-TRAP.
DIONÆA MUSCIPULA. Lind. 433.
White, middle of trap pink. 6 in.
Carolina. July—Aug.

TAMARICACEÆ—TAMARISK.

French Tamarisk is a slender shrub growing on the cliffs on the Southern coast.

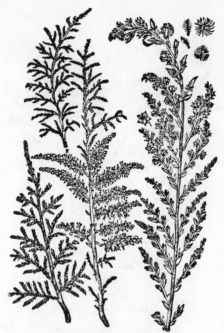

150.—FRENCH TAMARISK.
TAMARIX GALLICA. Z. 201.2.
Pale pink. 4-8 ft. July.

CARYOPHYLLACEÆ—PINKS.

Herbs with opposite undivided leaves set at swelling nodes, or "joints," in the stem. Flowers with 5 petals, 10 stamens, and 2, 3, 4, or 5 styles. The calyx is either of 5 sepals, or is a tube opening at the top by five points. Those plants which have sepals are called the Alsineæ; those with the tubular calyxes are the Sileneæ. The tubular calyx varies, and indicates the genus. In Dianthus and Saponaria it is smooth and unribbed. In Silene it is inflated and somewhat hairy and furrowed. In Lychnis, though bulging, it is hardly inflated, but is of firmer outline and ribbed. In Dianthus there is an epicalyx, or involucre, of bracts at the base of the tube.

The petals are usually of graceful form with slender claws attaching them to the flower. They are generally notched, sometimes so deeply as to seem to be 10 instead of 5. The outer edge, besides being notched, is sometimes toothed and cut,—hence particularly the name "feathered Pink." There is a tendency also for projections or shoulders to form one on either side of the petal, see the small figure below Fig. 166. In Ragged Robin these side projections are long, and with the long strap-like points made by the deep notch make each petal of 4 points or straps. This form is not correctly shown, I think, in any of the old books.

Where the petals become narrow and form the eye of the flower there are in some species white scales, closing the throat of the flower.

The inflorescence is definite, and often forks regularly; that is, the stem ends in a single flower, below which two branches go off either way. These bear a terminal flower, and fork in the same way. The flower standing in the fork is the first to develop. Even in the dense corymbose heads of Sweet William this same forking will, on inspection, be found to be present.

The styles are simple and tapering, the inner surface being stigmatic. The two curled styles in Pinks form notable decorative features.

Gypsophila, a genus of very delicate plants, from South Europe, belongs to this Order.

See in the Appendix—Fig. 26, Superb Pink; Fig. 25, Carnations; and Fig. 56, Scarlet Lychnis.

Fig. III. in the Frontispiece was evidently copied from Fig. 152 opposite.

151.—CARTHUSIAN PINK.
DIANTHUS CARTHUSIANORUM. F. 352.
Red. 18 in. July—Aug.

152.—FEATHERED PINK.
DIANTHUS PLUMARIUS. F. 353.
White. 6 in. Europe. July—Sept.

GARDEN PINKS are chiefly derived from D. plumarius, D. deltoides, D. Armeria, and D. Carthusianorum. They have deeply-cut petals, covered at the base with a soft whitish down. MAIDEN PINK, D. deltoides, has a rose-coloured flower, spotted with white, with a white eye enclosed in a crimson ring. It is a British species.

153.—CARNATION. CLOVE GILLIFLOWER.
DIANTHUS CARYOPHYLLUS. F. 354.

In the single wild state it is the CASTLE PINK, and grows on old walls. Its colour is flesh-colour, or crimson ;
and it flowers in June and July. The cultivated Carnations are derived from it. PICOTEES are such as are
white or yellow with pink, red or purple spots, or have the edges of one of these colours. FLAKES and
BIZARRES have striped petals—the stripes widening to the margin—and, in the former, of one colour upon white,
in the latter of two.

154.—SUPERB PINK.
DIANTHUS SUPERBUS. L. i. 451.1.
White. 2 ft. July—Sep.

155.—CARNATION.
DIANTHUS CARYOPHYLLUS. J. 588.1.
Crimson. 2 ft. June—Sep.

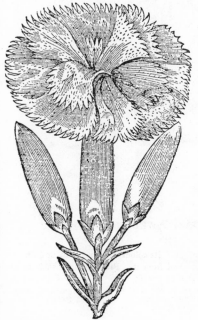

156.—GREAT DOUBLE CARNATION.
G. 472.1.

157.—"THE WHITE CARNATION
AND PAGEANT." J. 589.1.

158.—DEPTFORD PINK.
DIANTHUS ARMERIA? G. 478.1.
Pink, white dots. 1 ft. July—Sep.

159.—SWEET WILLIAM.
DIANTHUS BARBATUS. G. 479.3.
Crimson, pink, or white in rings.
18 in. June—July.

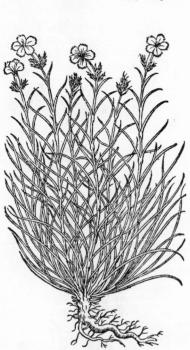

160.—"WHITE MOUNTAIN PINK."
DIANTHUS CÆSIUS? G. 476.9.
Flowers, rose, fragrant. Leaves glaucous.

161.—SOAPWORT.
SAPONARIA OFFICINALIS. Z. 1024.1.
Pink. 2 ft. Fl., 1 in. Aug.

162.—SOAPWORT?
M. 449.1.

163.—DOUBLE SPATLING POPPY.
?SILENE CUCUBALUS. G. 551.3.

164.—SEA CAMPION.
SILENE MARITIMA.
9 in. Fl., 1 in. June—Sep.
Petals white ; calyx pale umber, with
umber or purple lines. Grows in a dense
patch, on rocks close by the sea.

165.—" WHITE WILD CAMPION."
G. 384.8.

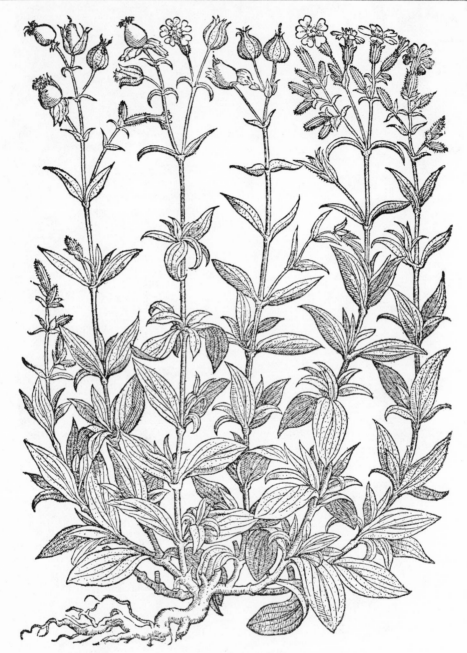

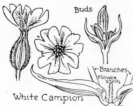

Buds

Branches
Flower
stem.

White Campion

166.—WHITE CAMPION.

LYCHNIS DIOICA VESPERTINA. M. (1565) 997.

Lychnis dioica is both white and red. The white opens in the evening and is fragrant. 2 ft. Fl., 1 in. June—Sep.

114

167.—"RED WILD CAMPION."

Lychnis dioica diurna. G. 382.1.

The Red Campion opens in the day-time. 2 ft. Fl., 1 in., the petals narrower than in the White. May—June.

168.—RAGGED ROBIN.
LYCHNIS FLOS-CUCULI.　G. 480.1.
Rose.　2 ft.　Fl., 1 in.　June—Aug.　Pet. incorrect.

169.—RED BACHELOR'S BUTTONS.
LYCHNIS DIOICA.　L. i. 336.1.

170.—WHITE BACHELOR'S BUTTONS.
LYCHNIS DIOICA.　L. i. 336.2.

171.—CORN COCKLE.
AGROSTEMMA GITHAGO.　F. (1545).
Purple.　3 ft.　Fl., 1½ in.　June—July.

172.—ROSE CAMPION.
AGROSTEMMA CORONARIA.　G. 381.1.
3 ft.　Flowers crimson, 1½ in.

173.—ROSE CAMPION.
AGROSTEMMA CORONARIA.　Z. 1024.2.
Leaves pale blue-grey.　June—Sep.

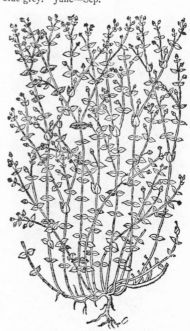

The ALSINEÆ are the plants with separate sepals.　They include the CHICKWEEDS (of which STITCHWORT, Holostea, is the handsomest), PEARLWORT, Sagina, SANDWORT, Arenaria, CORN MOUSE-EAR, Cerastium, and SPURREY, Spergula.

The petals are in some quite unnotched, in others deeply notched.　In Spurrey the leaves are in whorls.　The figure of it (Fig. 178) does not very well represent its flower-branches, which branch more horizontally, bear the flowers at greater intervals on fairly long pedicels, which hang down in fruit.

174.—THYME-LEAVED SANDWORT.
ARENARIA SERPYLLIFOLIA.　G. 488.3.
White.　4 in. common.　May—July.　Br.

175.—STITCHWORT.
STELLARIA HOLOSTEA. C.E. 743.

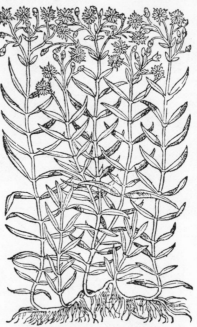

176.—STITCHWORT. STARWORT.
STELLARIA HOLOSTEA. G. 43.1.
White. 1 ft. Fl., ⅝ in. May—June.

177.—CORN MOUSE-EAR.
CERASTIUM ARVENSE. G. 477.11.
White. 5 in. June—July.

178.—SPURREY. PICK-PURSE.
SPERGULA ARVENSIS. D. 1331.2.
White. 1 ft. June—Sep.

MALVACEÆ—MALLOW.

Herbs and Shrubs. The flowers have a double calyx. The outer, or epicalyx, or involucre, is of 3 bracts in the genus Malva, of 3 bracts in Lavatera, and of 6 to 9 in Althæa. The petals are 5, heart-shaped, or, rather, fishtail-shaped. They are often streaked with 3 or 5 veins, usually crimson. The petals are mauve (which is the French name for the Mallow), but sometimes purple, rich crimson, rose, white, or pale yellow. Sometimes the depth of the flower is deeply coloured, blending to white at the edges of the petals—as in Hibiscus, and more or less in the Hollyhocks.

The stamens are united by their filaments into a tube, or column, through which the style passes. The style ends in about 20 bristle-shaped stigmas.

The buds often have a form very like the pointed polygonal domes of the East. The fruit is somewhat melon-like, sitting within the calyx, as is shown in Fig. 186.

The leaves are alternate, palmate, lobed and toothed, upon long stalks, and have stipules.

The British species are COMMON MALLOW; MUSK MALLOW; DWARF MALLOW, M. rotundifolia, which has its leaves but little indented; TREE MALLOW, Lavatera arborea, a biennial 3 to 12 ft. high, much like Common Mallow, but with its flowers darker towards the middle; and MARSH MALLOW.

The whole Order is most useful, but its chief plant in this respect is the COTTON-PLANT, Gossypium.

A fine engraving of the HOLLYHOCK is in the Appendix, Fig. 27.

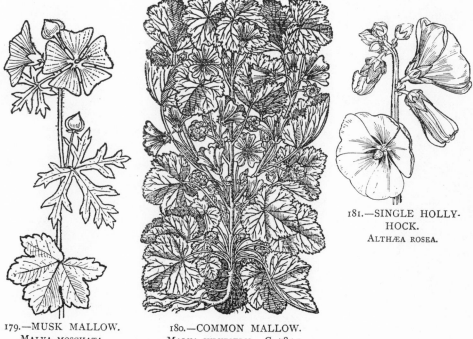

181.—SINGLE HOLLY-
HOCK.
ALTHÆA ROSEA.

179.—MUSK MALLOW.
MALVA MOSCHATA.
Rose. 2 ft. Fl., 1¼ in. July—Aug.

180.—COMMON MALLOW.
MALVA SYLVESTRIS. G. 785.1.

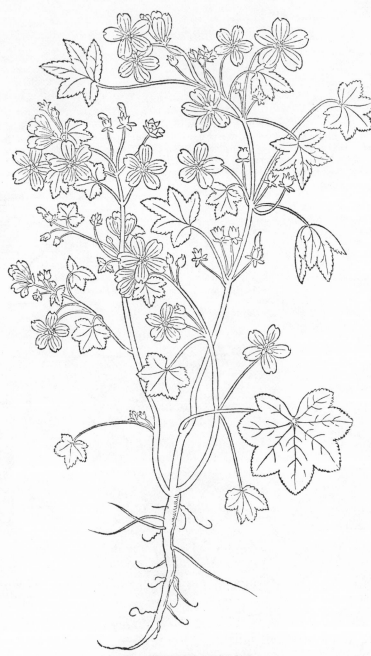

182.—COMMON MALLOW.

MALVA SYLVESTRIS. F. 509.

Flowers mauve. Epicalyx of 3 leaves. 3 ft. Fl., 1¼ in. May—Aug.
Common in hedges.

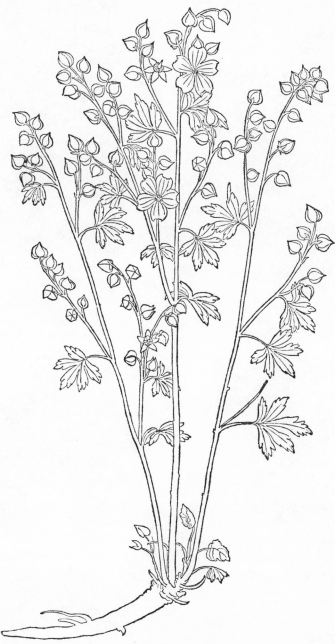

183.—HERB SIMON. VERVAIN MALLOW.
MALVA ALCEA. F. 80.
3 ft. Purple flowers. Native of Germany and Hungary.

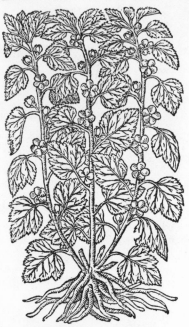

184.—MARSH MALLOW.
ALTHÆA OFFICINALIS. G. 787.1.
Rose.   5 ft.   Fl., 1 in.   July—Sep.

185.—HOLLYHOCK.
ALTHÆA ROSEA. G. 783.4.
White, Rose, etc.   8 ft.   Fl., 3 in.   Aug.—Sep.

186.—"WATER MALLOW."
"ALTHÆA PALUSTRIS." J. 933.2.
Probably this is an Abutilon.

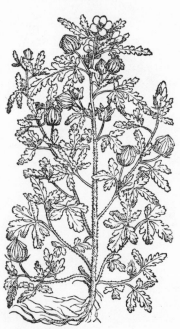

187.—VENICE MALLOW.   GOOD-NIGHT-AT-NOON.
HIBISCUS VESICARIUS. M. 472.
Yellow, with brown lines.

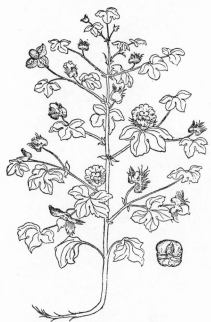

188.—COTTON [Malvaceæ].
Gossypium Herbaceum.　F. 581.
Yellow.　3 ft.　E. Indies.　July.

CAMELLIACEÆ—CAMELLIA, TEA.

An Order of Shrubs and Trees, with alternate leathery leaves, usually without stipules. Flower-stalks, terminal or axillary. Flowers generally white, seldom pink or red. Sepals 5 or 7. Petals 5, 6 or 9. The most noteworthy plants are the species of Thea, Tea. The Mangosteen, the fruit of Garcinia Mangostana, is reputed the most delicious of all fruit. The different varieties of Camellia japonica " are the glory of gardeners." They are shrubs about 10 ft. high, flowering in May to July. The flowers are 2½ in. across; they are red, pink or white—the white sometimes have a medial line of red.

189.—CAMELLIA JAPONICA.

TILIACEÆ—LIME.

The Tiliaceæ include herbs and shrubs, but none are British, and the Order is known to us only in the Lime Trees, of which there are three species in Britain, distinguished by the relative length of leaf-stalk to leaf. The flowers have 4 or 5 sepals and petals, numerous stamens, and 1 style, with as many stigmas as there are carpels in the fruit. The petals are lemon-yellow, very fragrant, and produce much honey. The panicles of flowers hang down—not as drawn on the next page. The long bract adhering to the flowering-stalk is correctly shown. Flowers ⅝ in. June—July.

123

190.—THE LIME TREE.
TILIA EUROPÆA. M. (1565) 174.

Bright
Lemon
Yellow
Petals

HYPERICACEÆ—ST. JOHN'S WORT.

These are mostly undershrubs, some of which are evergreen (H. calycinum), but some are herbs. The leaves are generally opposite. The flowers are yellow, with 4 to 5 unequal sepals, and 4 to 5 petals, sometimes unsymmetrical. Stamens numerous, in 3 or 5 bundles. Styles, 3 or 5. There are 12 British species.

H. perforatum was formerly in extensive repute as a charm against witchcraft and enchantment. It was gathered for the purpose on St. John's Day.

A square-stemmed St. John's Wort was called St. Peter's Wort.

Gerarde gives instructions for making a balsam for the cure of deep wounds and such as were "thorow the body."

191.—ST. JOHN'S WORT.
Col. 78.
Much enlarged from the etching.

192.—LARGE-FLOWERED ST. JOHN'S WORT.
HYPERICUM CALYCINUM.
Yellow. 18 in. Fl., 2 in. July—Sep.

193.—MOUNTAIN ST. JOHN'S WORT.
HYPERICUM MONTANUM. F. 76.
Pale Yellow. 2 ft. Fl., ¾ in. July.

194.—TUTSAN.
HYPERICUM ANDROSÆMUM. G. 435.
Yellow. 3 ft. Fl., ¾ in. July—Aug.

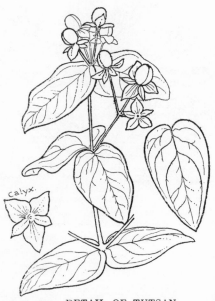

195.—DETAIL OF TUTSAN.

196.—COMMON ST. JOHN'S WORT.
HYPERICUM PERFORATUM. G. 432.1.
Yellow. 2 ft. Fl., ¾ in. July—Aug. Britain.
This figure is enlarged. It forms a good panel when coloured, but, like Fig. 191, is disappointing
on the larger scale.

SAPINDACEÆ—HORSE CHESTNUT.

Usually large trees, amongst which is the HORSE CHESTNUT—Æsculus Hippocastanum—an Asian tree introduced into England early in the seventeenth century. Soapy properties are possessed by some species of Sapindus, the fruits of which are said to be used as soap in the West Indies. The fruit of the Horse Chestnut is said to share this property.

The HORSE CHESTNUT is of rapid growth, and has indeed been regarded as a weed—Pope was blamed on that account for admitting it to his garden. It will attain the height of 60 ft., forming a parabolic head. Its leaves are very large, opposite, upon long stalks. The flowers are in terminal panicled racemes. Several flowers are upon each of the short flower-stalks which branch off from the upright peduncle. The petals are 5, white, or pink, some of the white being marked with crimson. There are 7 stamens, drooping gracefully. Style 1. The tree blooms in May. The fruit is similar in form to that of the sweet Chestnut. It is used as food for sheep, and starch is derived from it. It is enclosed within a burr.

197.—HORSE CHESTNUT.
ÆSCULUS HIPPOCASTANUM. D. 33.1.

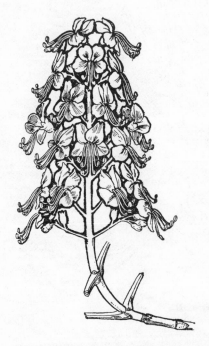

198.—FLOWER OF HORSE CHESTNUT.

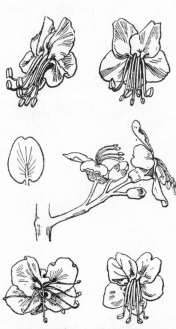

199.—DETAILS OF FLOWERS.

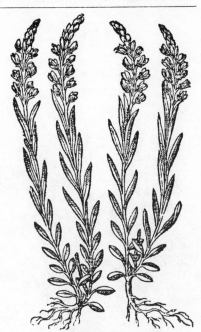

POLYGALACEÆ—MILKWORT.

Sepals 5, the inner 2 generally large and petaloid, wing-like, coloured. Petals combined by their claws to the filaments of the stamens. The lowest petal keeled and crested.

COMMON MILKWORT—Polygala vulgaris. 8 in. Blue, purple, pink, or white. Dry hilly pastures. June—July.

200.—COMMON MILKWORT.
G. 449.4.

ACERACEÆ—MAPLE.

The two common species of this Order are the SYCAMORE (called in the North, PLANE-TREE) and the COMMON MAPLE. The leaves of the different species of Acer vary very considerably, but the shape of those of these two common trees is well known. The most striking difference between them is in the position of the flower-branch. In Sycamore it hangs down, and is a raceme ; in Maple it is a branched panicle, and holds itself upwards. The difference in the spreading of the wings of the fruit is well shown in the cuts.

The Sycamore is remarkable for its very upright growth.

(For the true Plane-tree see Platanus, p. 402.)

201.—NORWAY MAPLE.
ACER PLATANOIDES.
A. Leaf of COMMON MAPLE.

203.—MAPLE.
ACER CAMPESTRE.　J. 1484.2.

202.—MAPLE.
ACER CAMPESTRE.　Bock. ii. 27.

204.—SYCAMORE.
ACER PSEUDO-PLATANUS.　J. 1484.1.

AURANTIACEÆ—CITRONS.

Trees or shrubs of evergreen character, having alternate leaves without stipules, with the blade articulated to the petiole, which is sometimes winged. Flowers fragrant, terminal or axillary, with a short calyx of 3 to 5 points; petals the same in number, white; stamens as many, or some multiple; style simple, stigma enlarged and somewhat divided. Uncultivated plants have sometimes axillary spines. The leaves are toothed, or entire.

The principal genus is Citrus, and includes the Orange, Lemon, Lime, Citron, Shaddock and Grape-fruits. The fruit differs much in shape and size. The Orange is generally globose, the Lemon ovoid, the Citron ovoid and rugose, the Shaddock spheroidal, and broader at the outer end, and very large. The Grape-fruit is a smaller kind of Shaddock, but still larger than the Orange, and grows in clusters. The Lime has but small fruit, $1\frac{1}{2}$ inches across, with a smooth, greenish-yellow rind, and has a pointed tip like the Lemon. All these mentioned are 15 feet high, except Lime and Citron, which are 8 feet. They flower from May to July.

These plants are natives of Asia, particularly of Eastern Asia. The Citron was cultivated in Italy in the second century, the Orange not till the twelfth or fourteenth. The "Golden Apples" of the ancients, and the "Forbidden Fruit" of the Jews, are supposed to have been some of these.

205.—ORANGE.
CITRUS AURANTIUM. G. 1279.3.

206.—LEMON.
CITRUS LIMONUM. G. 1278.2.
White. 15 ft. May—July.

207.—ORANGE.
CITRUS AURANTIUM.　M. (1565) 245.
15 ft.　Bark, greenish-brown.　Flowers, white.　May—July.

[The signature W. S. on the lowest leaf is, apparently, the only indication of the authorship of these large drawings from Matthiolus.]

208.—CITRON.
Citrus medica. G. 1278.1.
8 eet. Flowers, white. May—July.

[This figure is enlarged from the same size as Figs. 205 and 206, which are about the same size as the original woodcuts in Gerarde and Tabernæmontanus.]

VITACEÆ—VINE.

Climbing shrubs, with separable joints.
Leaves, with, or without, stipules, opposite
below, alternate above. Flower stalks, racemose, often opposite the leaves, and sometimes
by abortion converted into tendrils. Flowers
small, green, with 4 to 5 petals, soon falling,
and in the VINE lifted off in a mass by the
stamens, which are opposite to them and of
the same number.

RAISINS and CURRANTS (Corants—grapes
of Corinth) are the dried fruit of certain
varieties.

VIRGINIAN CREEPER (Ampelopsis quin-
quefolia) is a native of North America, where
also is the FOX GRAPE (Vitis vulpina).

The common Vine rises to 30 ft., and
flowers in June and July.

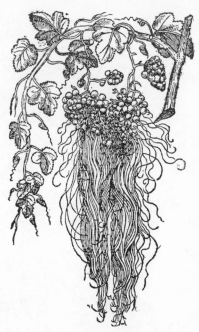

209.—"LACED OR BEARDED GRAPES."
G. 726.3.

210.—"STARVED OR HARD GRAPES."
G. 727.5.

211.—VINE.
Z. 294.

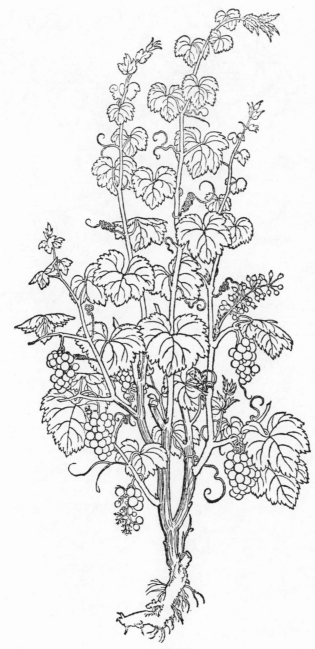

212.—VINE.

VITIS VINIFERA.   F. 84.

213.—VINE.

VITIS VINIFERA.   M. (1565) 1320.

214.—VINE.

VITIS VINIFERA. G. 725.1.

Nowhere is the tendency to arrange the plant in a decorative manner more evident than in this figure, which is one of the cuts of Tabernæmontanus, used in Gerarde. It bears some resemblance to the figure in Bock, and is very likely based upon that design. In Bock's figure there is an upright stick for the plant to twine upon, and at its foot is represented Noah, drunken, with his sons jeering at him.

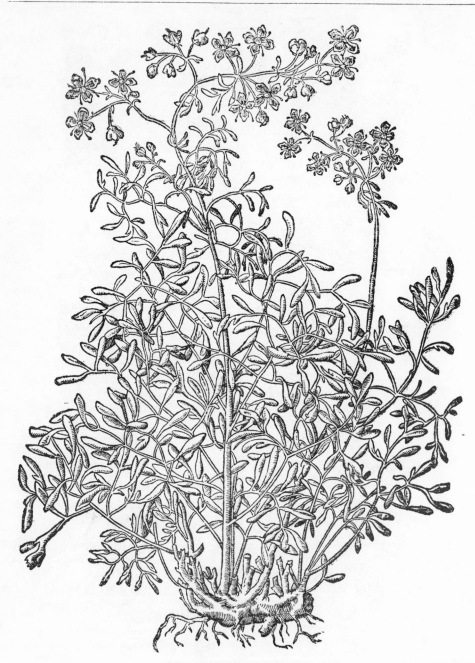

215.—RUE. HERB GRACE.

RUTA GRAVEOLENS. M. (1565) 737.

3 ft.   Petals yellow-green, deeply concave.   South Europe.   Gardens.   June—Sep.

RUTACEÆ—RUE.

Exotic plants with flowers with 4 or 5 sepals and petals, and fruit consisting of several carpels, more or less cohering. Stamens as many, or twice as many, as the petals. Style 1. The Order is commonly represented by RUE and DITTANY.

DITTANY is a garden flower from Switzerland. When rubbed it smells like lemon-peel, and it emits a vapour which is said may be seen to take fire, if a light be held near it—"a blue flame running over every part of the plant," and visible even in twilight.

216.—DITTANY.　BURNING BUSH.
DICTAMNUS FRAXINELLA.　J. 1245.
Pale rose-purple, with crimson lines.　3 ft.
May—July.

ZYGOPHYLLACEÆ—BEAN CAPER.

An Order of herbs, shrubs, or trees having opposite leaves, with stipules, and flowers with their parts in 4's or 5's. The Order includes GUAIACUM and BEAN CAPER—Zygophyllum Fabago. Syrian Rue, of which a graceful cut is here given, produces seeds which are used as spice.

217.—SYRIAN RUE.
PEGANUM HARMALA.　C.E. 496.
White.　1 ft.　Spain.　July—Aug.

LINACEÆ—FLAX.

Herbs, or even small shrubs, with simple entire leaves, which are either alternate, opposite, or whorled, and with showy, but fugacious, flowers. Sepals, petals and stamens usually 5, sometimes 4 or 3. The styles generally the same. Petals blue, red, yellow or white. Stamens joined together below.

Linum usitatissimum is one of the most useful of plants. The stems supply flax, and its seeds linseed-oil.

The handsome figures from Gerarde are not distinguishable in species, except perhaps the narrow-leafed. None show the petals slightly notched as they are in L. usitatissimum. There are many fine garden species.

218.—WILD FLAX.
G. 446.2.

219.—THIN-LEAFED WILD FLAX.
G. 446.1.

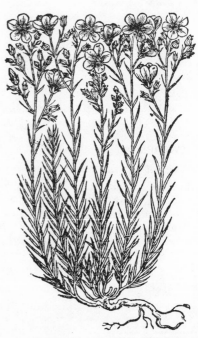

220.—THIN-LEAFED WILD FLAX.
LINUM ANGUSTIFOLIUM. G. 447.3.
Pale blue. 1 ft. Fl., ½ in. July.

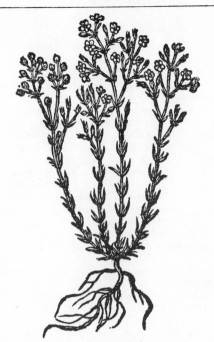

221.—WILD DWARF FLAX.
J. 559.4.

222.—ALL SEED.
RADIOLA MILLEGRANA. J. 569.2.
White. 2 in. July—Aug.

GERANIACEÆ—CRANE'S-BILLS.

This Order is sometimes divided into four—Geraniaceæ, Oxalidaceæ, Balsaminaceæ, and Tropæolaceæ. Geraniaceæ include *Erodium, Geranium, Pelargonium* and *Monsonia.* Oxalidaceæ are the WOOD SORRELS— *Oxalis.* Balsaminaceæ are the TOUCH-ME-NOTS—*Impatiens;* and Tropæolaceæ are the INDIAN CRESSES, commonly called NASTURTIUMS. The geraniums of the garden and of the cottage-window are pelargoniums. There is little external similarity between Touch-me-not and the common geraniums, except in the springiness of their fruits and the redness of their stems, stalks, and leaf-veins. Indian Cresses, which include CANARY CREEPER, have a growth, and a smoothness and juiciness, which do not suggest their connection with geraniums, but their form reveals their relationship. They climb by their twisting leaf-stalks.

The Geraniaceæ have opposite or alternate leaves, usually upon long, downy or hairy stalks, which are often swollen at the base, and are readily separable from the stem. There are stipules, particularly noticeable in Pelargonium, but absent in the Oxalids, Balsams and Indian Cresses. When the leaves are opposite, the flower-stalks rise from the axils, but when the leaves are alternate the flower-stalks are opposite to them. But this is not the case in the Indian Cresses, in which the leaves are alternate and the flowers axillary.

In form the leaves are either palmate or pinnate. In both cases they are generally lobed and toothed. The palmate leaves are usually circular or pentagonal in outline. They are obscurely lobed in Indian Cress, but deeply lobed and toothed in many other plants, often assuming an appearance similar to the leaves of Musk Mallow. In many Pelargoniums the leaves are variegated in a concentric manner. Hence some "geraniums" bear a "horse-shoe," and others have pale or whitish edges. The general tendency in the Order to redness of stem and leaf further beautifies the plant—the leaves sometimes being red at the edges, and turning red as they die. Many of the plants have a pleasant musky scent.

The flowers are commonly on somewhat short pedicels, 2 or more together at the summit of a stout peduncle, where are a few bracts, sometimes membranous, pointed, and reddish. The calyx is of 5 sepals, sometimes united together. Sometimes a spur is formed as in Tropæolum. In Pelargonium there is a spur, but it is connate to the stalk, and therefore unobserved. The calyx is green, but sometimes coloured—in Tropæolum it is yellow or orange.

The corolla is of 5 petals, often heart-shaped, and usually with 2 to 5 coloured or pellucid veins. The petals are red, violet, purple or white, and are sometimes splashed with colour, usually with red. This colouring occurs particularly on the 2 upper petals in Pelargonium and Tropæolum. In Tropæolum the flowers are yellow, orange or red; and one of the Balsams, Impatiens fulva, is yellow, and so is one of the Wood Sorrels, Oxalis corniculata.

The petals are irregular in *Pelargonium*, the upper 2 being smaller or larger than the lower 3. In Tropæolum they are irregular, and also fringed. In the Balsams they are very irregular and confused with the calyx. In *Monsonia* the petals have several teeth at the outer edge, after the manner of Pinks.

The stamens are normally 10,—15 in Monsonia. Five are fertile in Erodium, 7 in Pelargonium, and 10 in Geranium.

The fruit in the Crane's-Bills proper is remarkable for the long beak, or torus, around which, at the base, are 5 cells, each with its style, which, at first adhering to the beak, afterwards separates from it, from below upwards, carrying its cell with it. At the summit of the beak are 5 stigmas provided by the 5 styles.

The fruits of the Oxalids, Balsams and Indian Cresses are not of this character. The 3-celled fruit of the Common Indian Cress or Nasturtium is used as a substitute for Capers.

Common Wood Sorrel—Oxalis Acetosella—is one of the plants put down as the true Shamrock, though Trifolium minus appears to have a better claim.

The Pelargoniums are mostly from the Cape of Good Hope, and none seem to have been in cultivation here before 1724. Of Oxalids the Cape also furnishes many species; in Britain there are but 2—the White and the Yellow.

The names of the commonest genera in this Order refer to the resemblance of the fruit to the head of a long-billed bird. Geranos is Greek for a crane, Pelargos for a stork, and Erōdios for a heron.

There are 13 wild species of Geranium in Britain, 3 of Erodium, 2 of Oxalis and 2 of Impatiens. Yellow Wood-Sorrel, Oxalis corniculata, has alternate leaves and much the growth of a geranium. There are many exotic species of Oxalis. Details are given of one kind, Fig. 224. In it, as in others, the stems rise and bear umbels of leaves and flowers, whereas in our White Wood-Sorrel there is no stem.

The most common of the Balsams has red shiny stems and red-veined leaves, and flowers with its spurred petal deep and pitcher-like, and its petals pink, spotted with red.

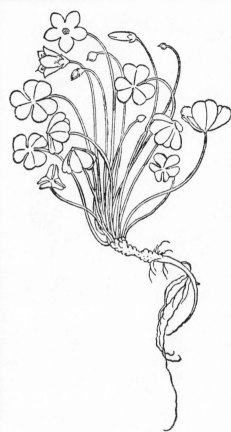

223.—WOOD SORREL.
Oxalis Acetosella. F. 564.
White. 4 in. Fl., ⅝ in. Ap.—May.

225.—TOUCH-ME-NOT. YELLOW BALSAM.
Impatiens Noli-me-tangere.
Yellow, orange spots. 2 ft. Fl., 1½ in. July—Aug.

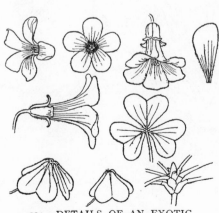

224.—DETAILS OF AN EXOTIC
SPECIES OF OXALIS.

225A.—INDIAN CRESS.
Tropæolum majus.
Orange, yellow. Fl., 2½ in. Peru. June–Oct.
"GERANIUM."
Pelargonium.
Red or white. 2 ft. Fl., 1 in. June–Sep.

226.—MEADOW CRANE'S-BILL.

Geranium pratense. M. (1565) 857.

1–2 ft. Flowers, 1¼ in., purple-blue with pellucid veins. Stamens dilated at the base. Stems swollen at the joints as shown. Moist pastures. June—July.

227.—MEADOW CRANE'S-BILL.
GERANIUM PRATENSE. G. 797.2.
Purple-blue. 2 ft. Fl., 1¼ in. June—July.

228.—BLOODY CRANE'S-BILL.
GERANIUM SANGUINEUM. G. 799.2.
Crimson-purple. 1 ft. Fl., 1¼ in. June—July.

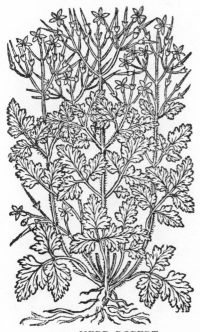

229.—HERB ROBERT.
GERANIUM ROBERTIANUM. G. 794.
Red, with white streaks. 18 in. Fl., ½ in.

230.—HERB ROBERT.
GERANIUM ROBERTIANUM.
Stems hairy, straggling. May—Sep.

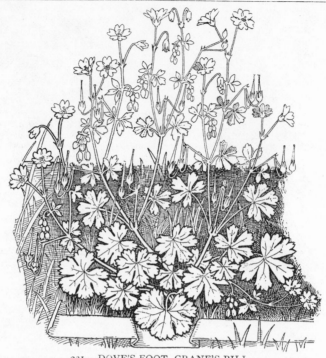

231.—DOVE'S-FOOT CRANE'S-BILL.
GERANIUM MOLLE.
Mauve. Petals notched. 6–18 in. Fl., ½ in. Dry pastures. Ap.—Aug.

232.—HEMLOCK HERON'S-BILL.
ERODIUM CICUTARIUM. G. 800.4.
Pink. 9 in. Fl., ½ in. or less. June—Sep.
Here shown in fruit.

233.—MUSK HERON'S-BILL.
ERODIUM MOSCHATUM. M. 402.1.
Pink. 6 in. Fl., ¼ in. June—July.

234.—SPINDLE-TREE. PRICKWOOD.

EUONYMUS EUROPÆUS. M. (1565) 191.

The CELASTRACEÆ are shrubs or small trees with inconspicuous flowers of greenish or white colour, and simple leaves. SPINDLE-TREE is the only British species. It is a hedge bush. Its flowers have 4 sepals and 4 greenish petals, but it is more remarkable for its deeply-lobed rose-coloured seed-vessels. Capsules, $\frac{1}{2}$ in. May. The wood is used for spindles and skewers.

[Here begin the Calycifloræ, with the stamens and petals upon the calyx.]

STAPHYLEACEÆ—BLADDER NUT.

Shrubs with opposite, rarely alternate leaves, and flowers with coloured calyxes—5-parted. Petals, 5, Stamens, 5.

235.—BLADDER NUT.
STAPHYLEA PINNATA. D. 102.
Pale greenish-yellow. Capsules green. June.

RHAMNACEÆ—BUCKTHORNS.

Trees or shrubs with simple alternate leaves with minute stipules, often thorny, and with inconspicuous flowers, and noted more for their berries.

Two species of Rhamnus are British ; and all species of that genus seem to be active as medicines, and to yield dyes and colouring matter.

The name of CHRIST'S THORN indicates that it is thought to have furnished the Crown of Thorns.

236.—CHRIST'S THORN.
PALIURUS ACULEATUS. Z. 260.2.
15–30 ft.

237.—THE JUJUBE-TREE.
Zizyphus vulgaris. M. (1565) 268.
Height 6 feet.

Zizyphus Lotus, which is similar to this plant, but has thorns at the bases of the leaves, shares with other plants, notably with Diospyrus Lotus, the honour of being the Lotus which furnished the fruit of oblivion to the Lotophagi of the *Odyssey*. It is a native of North Africa, and is 4 ft. high. Other plants said to have been the Lotus are Nitraria tridentata, and Celtis australis.

238.—COMMON BUCKTHORN.
RHAMNUS CATHARTICUS. M. (1565) 158.

COMMON BUCKTHORN has masses of 4-cleft, greenish flowers, with very narrow petals, succeeded by black berries. Its branches terminate in thorns. BLACK ALDER, which is the other British species, has 5-cleft flowers, whitish with purple anthers. The berries are purple-black, but at the same time some will be green, some red, and some black. ALATERNUS, R. Alaternus, is a tall shrub from South Europe with mellifluous green flowers without petals.

150

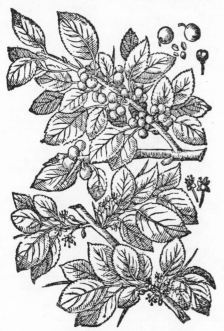

239.—COMMON BUCKTHORN.
RHAMNUS CATHARTICUS. Z. 261.
15 ft.　Berries, black.　Fl., May.

240.—BLACK ALDER.
RHAMNUS FRANGULA.　J. 1470.
12 ft.　Berries green, turning red and black in Aug.

ANACARDIACEÆ—SUMACH.

Shrubs and trees abounding in resinous, gummy juice. They are natives of South Europe, the Levant, North America, and Eastern Asia.　The leaves are alternate, and the flowers have usually 5 sepals, petals, and stamens. Pistacia Lentiscus yields Gum Mastic.　Rhus Vernix yields the true lacquer of Japan.　The Cashew Nut is the fruit of Anacardium occidentale.　The Turpentine Tree, or Terebinth, is Pistacia Terebinthus of South Europe.　Venetian Sumach is a shrub 10 ft. high.　Its leaves are of a delightful green, and emit a grateful scent when bruised.　The ends of last year's shoots divide, and produce hair-like bunches of purplish flowers, so as to cover the tree.　In the autumn these tufts remain, but of a darker colour.

241.—VENETIAN SUMACH. WIG TREE.
RHUS COTINUS. Z. 73.1.

151

LEGUMINOSÆ—PEAS.

This Order is divided into 3 Sub-Orders—

    1. Papilionaceæ,
    2. Cæsalpinieæ,
    3. Mimoseæ.

They have all, as a rule, a legume (familiar in the pea-pod) for fruit. The leaves are pinnate—in clover and other plants there is but one pair of lateral leaflets, so that the three leaflets make a trefoil—and frequently the terminal leaflet is wanting, and in its place is a short point, or a tendril, perhaps branched. There seems to be a tendency for the leaves to dwindle towards the point, and to be fuller at the base, where there are 2 stipules, sometimes narrow and inconspicuous as in Lupin, sometimes pointed, membranous and striated as in the Clovers, but sometimes large and leafy as in Peas. One species, Lathyrus Aphaca, has broad leafy stipules (Fig. 276), and no blade at all to its leaves, but only a mid-rib, branching at the end into tendrils.

The flowers differ in the 3 Sub-Orders. In PAPILIONACEÆ, which alone are native in Britain, the flower is butterfly-like (Fr. *papillon*, a butterfly). It has a calyx of one deep bell-shaped piece with 5 teeth at the margin, of which the odd one is anterior (away from the stalk). Often the 2 posterior teeth form together a back lip, and the 3 anterior a front lip, to the calyx, owing to two of the divisions being deeper than the others.

The corolla consists of 5 petals, of which 2 are conjoined and form the *keel*. Overlapping the edges of the keel, on either side, are 2 narrow petals called *wings*, and outside of these again, and much broader, is the fifth petal, the *banner* or *standard*. It is strongly creased down the middle, and, before expanding, is folded roof-wise over the rest of the petals.

The ovary—the future pod—is long and turns up at the outer end in a somewhat curved style—so noticeable on the pods of the Broom. The stamens are 10, and surround the ovary, 9 being joined together and a tenth being alone on the upper part. Hence the stamens are in 2 "brotherhoods," and the Papilionaceæ become the Diadelphia decandria of the Linnæan system. While this Sub-Order supplies some of the most valuable food-stuffs, it is, on the whole, poisonous, and the wholesome members of it are exceptional.

In the CÆSALPINIEÆ (after the genus Cæsalpinia—named in memory of Cæsalpinus) the petals are overlapped in precisely the opposite manner—the posterior petal (against the stem) being the innermost. The petals are more regular in general appearance. This Sub-Order is wholly exotic, and supplies medicines (such as senna), dyes, and edible fruits. "None possess any evident poisonous properties."—*Bentley.* Of the figures given, Pudding Pipe and Carob Tree belong to this Sub-Order.

In MIMOSEÆ the petals are regular, edge to edge, or are absent. The stamens are more numerous. ACACIA and MIMOSA are shrubs from 3 to 10 ft. high. Acacia vera yields Gum Arabic. Some species of Mimosa are Sensitive Plants. The flowers are in balls, and sometimes are like tassels.

The pods are in some cases of curious form. Horse-shoe Vetch—Hippocrepis Ferrum-equinum—has its seeds of horse-shoe shape and its pods notched accordingly. In other cases the pods are notched and twisted.

REST-HARROW is a trailing, rather shrubby plant. Its banner is striated with crimson lines—the wings are white. KIDNEY-VETCH has the flowers in woolly heads, two of which are close together upon a stalk—a large leafy bract beneath each. The genus Medicago is remarkable for its curved or spiral pods. Some pods so resemble snail-shells as for the plants to be called "snails." In the leaves of clover, let it be noted how parallel are the veins in each leaflet, and how they produce teeth on the edge. Also let it be noted that though the leaf-tip is often retuse or notched, there is a slight prolongation of the mid-rib. In Trifolium stellatum and Trifolium arvense the calyxes have very long teeth which produce in the first (which has rose-coloured petals and the calyx red at the base) a star-like effect, and in the latter the furry, "hare's-foot" character. The fruit of Ornithopus is a jointed legume—shown in Fig. 263.

LUPINS are of two kinds of growth—one kind branches into three, and into three again, with a spire of 2 or 3 whorls of flowers in the hollow between the three ascending branches and also at the branch-ends; the other kind sends up a tall spire—a raceme—perhaps 5 or 6 ft. high, with innumerable flowers spirally one beyond another. The banner in Lupin is rolled back at the two side edges so that it becomes narrow. In some kinds the banner is parti-coloured. The "Wild Ciche," as illustrated in Figs. 272 and 273, does not seem to be a near relation of the Chick Pea itself; it is rather a species of Ononis—Rest-Harrow.

Of the shrubs and trees the vast majority are exotic. The LABURNUM is from South Germany and Switzerland. The ROBINIA, or LOCUST TREE, or FALSE-ACACIA, of which there are many varieties, is from North America. The WISTARIA is from North America and China. These and many more are among the most beautiful shrubs and trees in our gardens.

The JUDAS TREE is 20 ft. high, produces rose-coloured flowers before the leaves, and is native to South Europe. "This is the tree whereon Judas did hang himself, and not upon the elder-tree, as it is said."—*Gerarde.*

Ceratonia Siliqua is generally considered to be the Locust-tree of Scripture. The beans are called Locust Beans, and in Spain they are called St. John's Bread, upon the supposition that he fed upon them. But it has been suggested (*vide* Loudon) that they were more likely the "husks" which the Prodigal Son was glad to eat.

The heart of the wood of Medicago arborea is dark in colour, hard as ebony, and used for sabre handles and monks' beads in the Levant.

GORSE, FURZE, or WHIN, Ulex europæus, is one of our commonest wayside shrubs. DWARF FURZE, Ulex nana, is similar, but smaller in every part.

See Appendix, Figs. 53 and 54, for Gorse, Judas-Tree, Acacia and Mimosa.

242.—WOODIE DYER'S WEED.
G. 1134.2.

243.—NEEDLE FURZE.
GENISTA ANGLICA.   F. 220.
Bright yellow.   2 ft.   May—July.

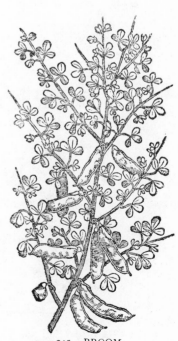

244.—BROOM.
CYTISUS SCOPARIUS.
Bright yellow.   8 ft.   Fl., 1 in.   May—June.

245.—BROOM.
CYTISUS SCOPARIUS.   M. 114.

246.—LABURNUM.
CYTISUS LABURNUM.
Yellow.　20 ft.　Fl., ¾ in.　S. Europe.　May—June.

247.—REST-HARROW.
ONONIS ARVENSIS.
Rose, with red veins ; wings white.
June—Sep.　Br.
This figure shows (full size) the summit
of one of the trailing shrubby, densely
leaved stems, perhaps a foot long.

248.—KIDNEY VETCH.
ANTHYLLIS VULNERAR　G. 1060.1.
Yellow, red or white.　9 in.　June—Aug.

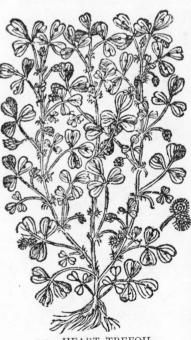

249.—HEART TREFOIL.
MEDICAGO MACULATA.　J. 1190.4.
Yellow.　8 in.　May—June.

250.—SHRUB TREFOIL.
? MEDICAGO ARBOREA. D. 259.1.

251.—RAY-PODDED MEDICK.
MEDICAGO RADIATA. L. ii. 38.2.
Yellow. 6 in. Italy. June—July.

252.—MOON TREFOIL.
MEDICAGO ARBOREA. L. ii. 46.2.
Yellow. 8 ft. Levant. May—Nov.

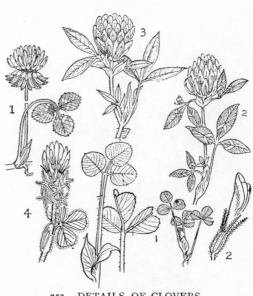

253.—DETAILS OF CLOVERS.
1. T. REPENS. 2. T. PRATENSE. 3. T. MEDIUM.
4. T. STELLATUM.

254.—DUTCH CLOVER.
Trifolium repens. L. ii. 29.1.
White. 3-9 in. May—Sep.

Of Clovers there are many kinds wild in Britain. Some are very small, and, though abundant, are easily overlooked. The best known are the cultivated species figured on this page—the White, the Red, and the Purple. What plant Trefoil of America is one cannot suggest. Treacle Clover was used as an antidote or counter-poison. A treacle was "a physical composition made of vipers and other ingredients" to the number of sixty! It was good for curing envenomed wounds, though if applied to wounds which were not envenomed it increased the pain by drawing to the wound "from far" those "humors" it must needs act against. The "Treacle Bible" has its name from its reading—"for there is no more Tryacle at Galaad."—Jer. viii. 22.

Hare's-foot Trefoil has its furry flower-heads whitish with the long feathery points of its calyxes. The corollas are too small to greatly influence the colour.

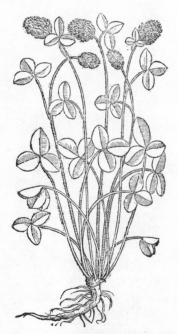

255.—RED CLOVER.
Trifolium pratense. M. (1554) 394.2.
Red-purple. 1 ft. July.

256.—GREAT PURPLE CLOVER.
Trifolium medium. M. (1554) 394.3.
Purple. 18 in. July.

257.—"TREFOIL OF AMERICA."
G. 1020.1.

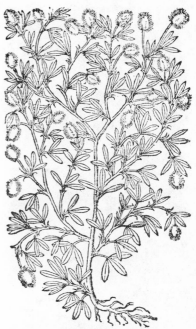

258.—HARE'S-FOOT TREFOIL.
TRIFOLIUM ARVENSE.　G. 1023.2.
Pale purple.　1 ft.　July—Aug.

259.—TREACLE CLOVER.
PSORALEA BITUMINOSA.　D. 504.2.
Pale blue.　4 ft.　S. Europe.　Ap.—Sep.

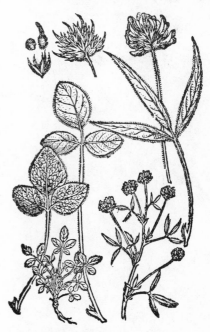

260.—TREACLE CLOVER.
PSORALEA BITUMINOSA.　Z. 899.
Pale blue.　4 ft.　S. Europe.　Ap.—Sep.

261.—BIRD'S-FOOT TREFOIL.
Lotus corniculatus. F. 527.

Varying in height from 3 to 8 in., and there is also a larger species, somewhat hairy, reaching to 2 ft. Flowers yellow, red in the bud. The stipules are very like the leaflets. This drawing is erroneous in not showing the leaves properly, nor the flowers in umbels. The next figure is correct, but omits the trefoil bract beneath the umbel. June—Sep.

262.—BIRD'S-FOOT TREFOIL.
LOTUS CORNICULATUS.　M. (1554) 504.1.
Yellow.　8 in.　June—Sep.

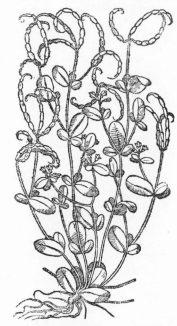

263.—PURSLANE-LEAFED BIRD'S-FOOT.
ORNITHOPUS SCORPIOIDES.　M. (1554) 562.
Violet.　6 in.　S. Europe.　June—July.

264.—COMMON MELILOT.
MELILOTUS OFFICINALIS.　Z. 830.2.
Yellow.　3 ft.　June—July.

265.—FENUGREK.
TRIGONELLA FŒNUM-GRÆCUM.　M. (1554) 232.
Yellow.　2 ft.　S. Europe.　June—Aug.

266.—"HATCHET VETCH."
ASTRAGALUS? M. (1554) 412.1.

267.—"HATCHET VETCH."
ASTRAGALUS. G. 1056.1.

268.—COMMON CATERPILLAR.
SCORPIURUS SULCATA. G. 267.
Yellow. 2 in. S. Europe. June—July.

FRENCH HONEYSUCKLE.—Hedysarum
coronarium is a handsome plant, 4 ft.
high, with scarlet flowers—from Italy.
SAINTFOIN, Onobrychis sativa, is also a
handsome plant, having a pyramidal head
of variegated crimson flowers. 2 ft. Br.

It will be noticed that there are here
represented herbs—from Clover and Medick
with trefoil leaves to Astragalus, Galega,
etc., with pinnate leaves. Several genera
and many species are unrepresented, both
native British and garden plants. These
genera vary chiefly in their fruit.

269.—GOAT'S RUE.
GALEGA OFFICINALIS. M. (1565) 739.
Flowers, blue or white. 4 ft. Spain. Gardens. June—Sep.

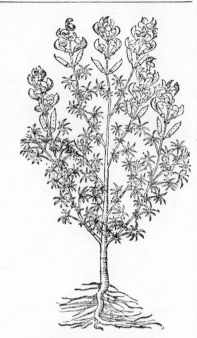

270.—YELLOW LUPIN (IN FRUIT).
LUPINUS LUTEUS. G. 1043.2.
Yellow. 2 ft. Sicily. July—Aug.

271.—SMALL BLUE LUPIN.
LUPINUS VARIUS. G. 1043.3.
Blue, white. 3 ft. S. Europe. July—Aug.

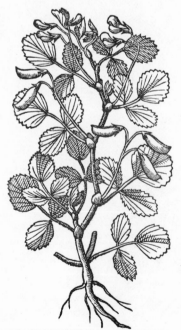

272.—WILD CICHE OR CHICK PEA.
L. ii. 73.1.

273.—WILD CICHE OR CHICK PEA.
D. 464.2.

Nos. 272 and 273 are probably some species of Ononis—Rest-Harrow.

Vetches and peas are botanically distinguished by the form and villous condition of the style. "LATHYRUS or EVERLASTING PEA has a flat style, villous above, growing broader upwards; in this it differs from the PEA, which has a triangular style keeled above: both genera have the two upper divisions of the calyx shorter than the other three, and in other respects are very nearly allied. Some species of LATHYRUS have one flower only on a peduncle: of these we have two wild ones—L. Aphaca and L. Nissolia. SWEET-SCENTED PEA—L. odoratus, with some few others, has two flowers on every peduncle; each tendril has a pair of oblong ovate leaves, and the legumes are rough. The banner is dark purple, the keel and wings light blue: but there are varieties; one all white, and another with a pink banner, wings of a pale blush, and a white keel; this is called PAINTED LADY PEA.

"VETCH or TARE is sufficiently distinguished by having a stigma transversely bearded on the under-side. The species, which are eighteen in number, may be ranged under two divisions—the first, with flowers in bunches on peduncles; the second, those which are axillary or have the flowers sitting almost close to the stem. Of the first division we have the TUFTED and WOOD VETCH—V. Cracca and V. sylvatica; to the second belong the CULTIVATED VETCH, V. Sativa, and BUSH VETCH, V. Sepium, which ramps in hedges, and has its leaflets entire and decreasing in size towards the end of the leaf."—*Martyn.*

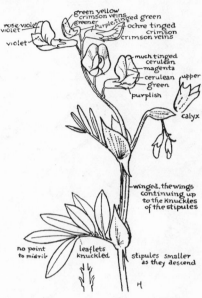

275.—TUBEROUS BITTER VETCH.
LATHYCUS MACRORRHIZUS.
Crimson fading to blue. 1 ft. May—July.

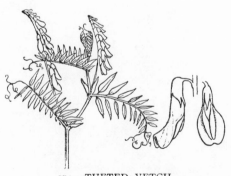

274.—TUFTED VETCH.
VICIA CRACCA.
Purplish blue. 3–5 ft. Fl., ⅝ in. July—Aug.

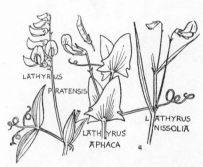

276.—MEADOW, YELLOW AND GRASS
VETCHLINGS.

LATHYRUS PRATENSIS. Yellow. 6 ft. July—Aug.
LATHYRUS APHACA. Yellow. 2 ft. June—Aug.
LATHYRUS NISSOLIA. Crimson. 2 ft. May—June.

277.—CHICKLING VETCH.
LATHYRUS SATIVUS. J. 1230.3.
Light blue. 3 ft. S. Europe. June—July.

278.—"EGYPTIAN CHICHLING."
J. 1230.4.

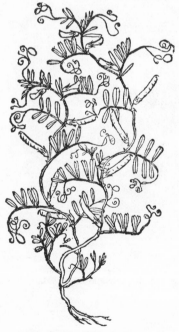

279.—COMMON VETCH.
VICIA SATIVA. J. 1227.4.
Purple. 3 ft. May—June.

280.—SMALL WILD TARE.
J. 1228.

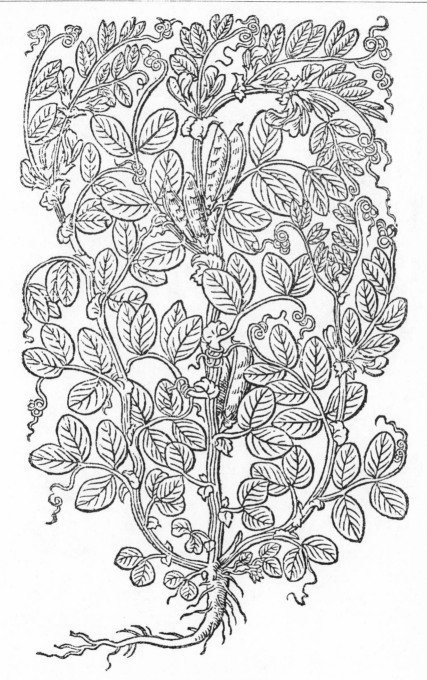

281.—THE WILD BEAN.

Vicia Faba. G. 1036.2.

Flowers purple, or white with a black spot on the wings.   3 ft.   Fields and gardens.   June—July.
This figure is enlarged from 5 in. high.

282.—BROAD BEAN.

Vicia Faba. M. (1565) 420.

Flowers purple, or white with a black spot on the wings. 3 ft. Fields and gardens. June—July.

The racemes are shown with too many flowers, which should be from 2 to 4. They were probably drawn from memory, when the plant was in fruit. The plant is here altogether too stunted. The execution of the drawing is very admirable, and the figure gives an excellent suggestion for modelling.

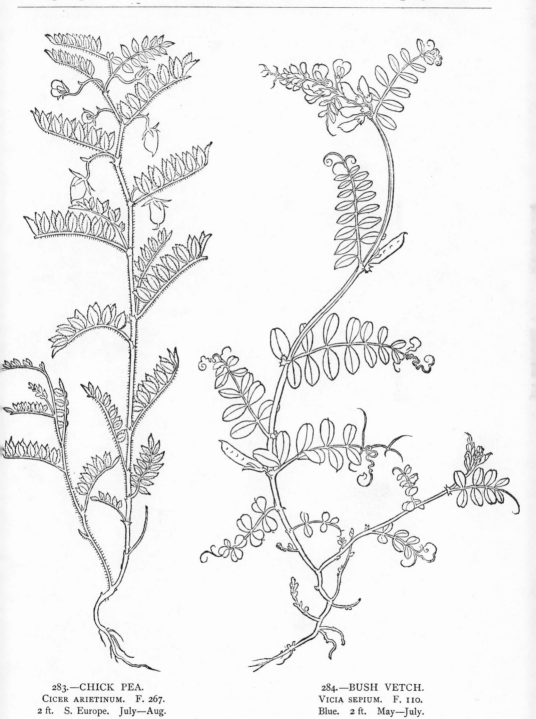

283.—CHICK PEA.
CICER ARIETINUM. F. 267.
2 ft. S. Europe. July—Aug.

284.—BUSH VETCH.
VICIA SEPIUM. F. 110.
Blue. 2 ft. May—July.

285.—EARTH-NUT PEA.
LATHYRUS TUBEROSUS. F. 131.
Flowers, red. 2 ft. Holland. The tubers are eaten. July—Aug.

286.—NARROW-LEAFED EVERLASTING PEA.
LATHYRUS SYLVESTRIS. F. 572.
Flowers, greenish and purple. 6 ft. July—Aug.

287.—LENTIL.

ERVUM LENS. F. 859.

Flowers, pale. 1 ft. Cultivated in France. May.

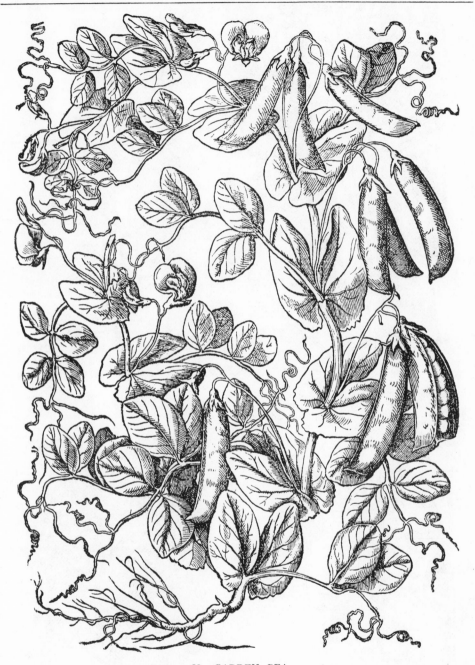

288.—GARDEN PEA.

Pisum sativum. M. (1565) 429.

Height, 2 to 12 ft., according to the variety—white or pink. Flowers one or more together. A plant whose origin is lost behind centuries of cultivation.

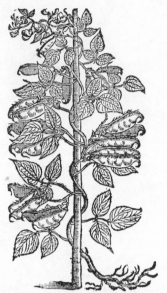

290.—"YELLOW KIDNEY BEAN."
PHASEOLUS. J. 1212.4.
The seeds are yellow.

289.—KIDNEY BEAN.
PHASEOLUS VULGARIS. M. (1554) 272.2.
White. 1 ft. From India. June—Sep.
The Scarlet Runner is Phaseolus multiflorus and rises to 12 ft.

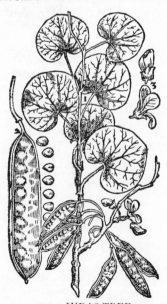

291.—JUDAS-TREE.
CERCIS SILIQUASTRUM. Z. 96.1.
Rose. 20 ft. S. Europe. May—June.
(See Appendix, Fig. 54.)

292.—BLADDER SENNA.
COLUTEA ARBORESCENS. G. 1116.1.
Yellow. 10 ft. France. June—Aug.

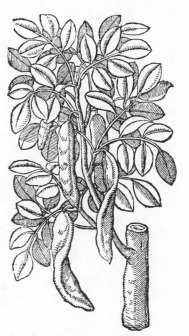

293.—PUDDING PIPE.
Cassia Fistula. L. ii. 104.2.
Yellow. 3 ft. E. Indies. June—July.

294.—CAROB TREE.
Ceratonia Siliqua. J. 1429.
Red. 15 ft. Levant. Sep.—Oct.

ROSACEÆ—PLUMS, APPLES AND ROSES.

This Order is, to the artist, the most important of all, from its including the Rose and the Apple.

The leaves are alternate, stalked, and with stipules. Their edges are serrated, and their general shape is (1) oblong-oval as in Apple, Pear, Cherry and Almond ; (2) pinnate, as in Rose, Mountain Ash, and Meadow-Sweet ; or (3) palmate, as in Lady's Mantle and Cinquefoil. In the pinnate leaves there are sometimes small leaflets between larger ones, while there is often a large lobe at the end as if three leaflets had become conjoined. In all kinds of leaves in this Order it is usual for the leaf to be wider towards the extremity. The leaves are commonly smooth and shiny above, whitish beneath.

The flower consists of a thickened calyx, usually with 5 points ; a corolla of, usually, 5 petals, which sometimes have very slender claws, are usually concave, and often notched ; 20 or more stamens, and 1, 2, 5 or more ovaries. In the number of stamens, pistils, petals, and points to the calyx, the few plants braced together as *Sanguisorbaceæ*—Lady's Mantle and Burnet—differ considerably from the other plants of the Order, and hardly seem to belong to it. The Linnæan Class, *Icosandria* (20 stamens), consists largely of plants of this Order, though it does not embrace all of them. The several "Orders" of the "Class" are *Monogynia*—Prunus ; *Digynia*—Cratægus ; *Pentagynia*—Mespilus, Pyrus, Spiræa ; *Polygynia*—Rosa, Rubus (both with 5-cleft calyx), Tormentilla, Dryas (both with 8-cleft calyx), Fragaria, Potentilla, Geum, Comarum (all with 10-cleft calyx—that is, a calyx of 5 points and an epicalyx of 5 more). These "Orders" indicate the number of styles.

The flowers are rarely solitary, but are generally in terminal clusters. The petals are sometimes broad and roundish, sometimes elongated and rather oblong, often notched, and adhere to the calyx by narrow claws. They are white, yellow, pink, or red, rarely purple and never blue. The numerous stamens are at first curled inwards—in the bud. This is well seen in the Common Laurel, in which the newly-opened flowers have all the stamens

173

with their heads bent inwards. Differences in the form of the fruit have caused the Order to be subdivided into certain Sub-Orders. The calyx is thick. Its lower part commonly becomes a receptacle upon or within which the fruit forms. The upper part of it, or its lip, is divided into 5 segments (4 in the Sanguisorbaceæ), and these five points remain upon, or under, the fruit in practically all plants but those of the Cherry Tribe. The main Sub-Orders are (1) *Pruneæ*, or Cherry Tribe ; (2) *Pomeæ*, or Apple Tribe ; and (3) *Roseæ*, or Rose Tribe. The *Roseæ* can again be divided into the *Rosæ*, the *Potentilleæ*, the *Spiræidæ*, and the *Sanguisorbaceæ*. In Pruneæ the fruit develops as a fleshy berry, with a "stone" in the middle, within, but not as part of the calyx, which falls away. In Pomeæ there are 5 carpels, or seed boxes, enveloped within the fleshy or pulpy receptacle, or lower part of the calyx. An apple cut across shows this. In Roseæ the fruit consists of many little nuts or pericarps arranged within, upon, or all over the receptacle, which is sometimes very hard, and sometimes (as in Strawberry) very soft. In Rosæ the receptacle is a hollow vase studded within with small nuts ; in Potentilleæ the nuts (which are sometimes like little plums—as in Blackberry) are studded over a somewhat globular receptacle ; in Spiræidæ the fruit consists of several separate follicles standing within the calyx ; and in Sanguisorbaceæ there are 1 or 2 nuts enclosed within a dry tubular receptacle.

It will be noticed that in the shrubs and trees of this Order some of the twigs are long, bearing perhaps 8 or 10 leaves upon them, and others are short, consisting apparently of a tuft of 4 or 5 leaves with a cluster of flowers rising from the middle. In the latter case a very short length of stem is produced, but in both instances the leaves are spirally alternate. The long twigs appear to grow in order to extend the boughs. COMMON LAUREL —Cerasus Laurocerasus—seems rarely, if ever, to develop tufts. Its boughs are generally horizontal, and its leaves also maintain a horizontal position instead of being spirally all round the twig. One is tempted to say that the normal manner of growth in this Order is for a wood-making twig of some length to be sent forth one year, and for there to appear upon it, in the next year, where the leaves have been, either a short knotty twig bearing a tuft of 4 or 5 leaves, and in the middle a cluster of flowers, or a short twig bearing leaves every few inches and terminating in a branched panicle, or a corymb, of flowers.

The flowers are nearly always in clusters at the end of the twig, and also rising from the axils of the last two leaves. Very often these three sets of flowers together form one head, a corymb. In HAWTHORN the boughs are richly clothed with flowers because all down the bough come forth little tufts of leaves each with its cluster of flowers.

In the Cherry Tribe generally, the flowers arise, singly or in umbels, from a scaly bud—appearing before the leaves : in CHERRY itself, when the leaves are young.

In the Almond and Peach the flowers are sessile. Double flowers are common. The calyx is deciduous—the fruit bears no vestige of it, and if it did it would be against the stalk, not, as in apple, at the eye of the fruit.

PEACHES grow in sheltered gardens, and are natives of Persia. The Peach is a downy fruit, the Nectarine a smooth one. They are said sometimes both to grow on the same branch.

Prunus domestica is probably a variety of P. insititia—WILD BULLACE TREE, and both, perhaps, of P. spinosa. The BULLACE TREE is a small hedge, or wood, tree, with black globular fruit with a fine bloom upon it. Its branches end in thorns, which in the cultivated plum have become absent. The DAMSON, P. damascena, is a variety of P. domestica, with small purple fruit. GREENGAGE is another variety, with green fruit. The dried fruit—PRUNE—is a dried plum.

BIRD CHERRY is so named because the birds like its fruit, and it is planted for their sake.

COMMON LAUREL must not be confused with the true Laurel or Sweet Bay. It is from the Levant, and makes a wide spreading mass of evergreen foliage.

PORTUGAL LAUREL, Cerasus lusitanica, is a similar plant classed as a low evergreen tree ; up to 30 ft. high in English gardens.

DWARF CHERRY is a bush which flowers and fruits most of the summer. Loudon, who calls it the Ground Cherry Tree, says that when grafted standard high it becomes a small round-headed tree with drooping branches.

The Pruneæ or Amygdaleæ—the Cherry Tribe—includes the Almond, the Peach, the Plum, the Cherry, the Sloe, and the Common Laurel and Portugal Laurel.

The Pomeæ—the Apple Tribe—includes the Apple, the Pear, the Quince, the Medlar, the Hawthorn, the Cotoneaster, the Mountain Ash, the Service, and the Beam Trees.

The Rosæ consist only of the genus Rosa.

The Potentilleæ—the Strawberry Tribe—include the Strawberry, the Blackberry, the Raspberry, the Avens, the Cinquefoils, and several similar small plants.

The Spiræidæ are mainly exotic shrubs, usually with pinkish flowers—they are represented in Britain by Meadow-Sweet and Filipendala.

The Sanguisorbaceæ comprise the Burnets and Lady's Mantle.

295.—ALMOND.
AMYGDALUS COMMUNIS. G. 1256.

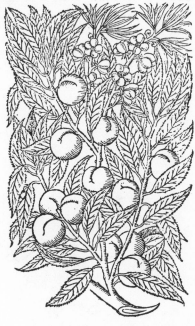

296.—PEACH.
AMYGDALUS PERSICA. G. 1258.4.

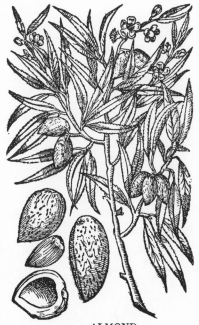

297.—ALMOND.
AMYGDALUS COMMUNIS. Z. 135.
Pale red, white. 20 ft. Fl., 1½ in. Mar.—Ap.

298.—PEACH.
AMYGDALUS PERSICA. C.E. 145.
Pink. 15 ft. Fl., 1½ in. Ap.—May.

299.—PEACH.

AMYGDALUS PERSICA. F. 601.

Flowers, $1\frac{1}{2}$ in., purplish-pink, sessile. Flowers sometimes double. Fruit, pale yellow, tinged rose.

15 ft. Persia. Sheltered gardens.

300.—CHERRY.

Prunus Cerasus. M. (1565) 233.

Flowers, 1 in., white, in umbels. 20 ft. Woods and orchards. May.

301.—CHERRY.
Prunus Cerasus. F. 425.

302.—BIRD CHERRY.
Prunus Padus. J. 1504.9.
White. 30 ft. Fruit, black. May.

303.—LATE-RIPE CHERRY.
G. 1320.5.

304.—"GASCONY CHERRY."
G. 1320.4.

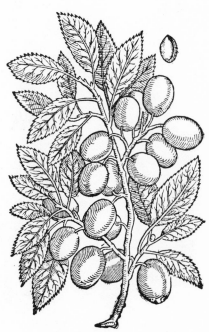

305.—YELLOW WAX PLUM.
"Pruna Cerea." C. E. 164.

306.—DWARF CHERRY.
Cerasus Chamæcerasus.   M. (1565) 236.
Flowers white.   8 ft.   Austria.   May.

307.—LESSER APRICOT-TREE.
PRUNUS ARMENIACA. D. 297.2.

308.—GREATER APRICOT-TREE.
PRUNUS ARMENIACA. G. 1260.1.
White. 15–30 ft. Feb.—Mar.

309.—DWARF CHERRY.
CERASUS CHAMÆCERASUS. D. 201.3.

310.—DAMSON.
PRUNUS DAMASCENA. G. 1312.4.
A variety of P. domestica.

311.—PLUM-TREE.

PRUNUS DOMESTICA. M. (1565) 265.

Flowers white. 20 ft. Ap.—May.

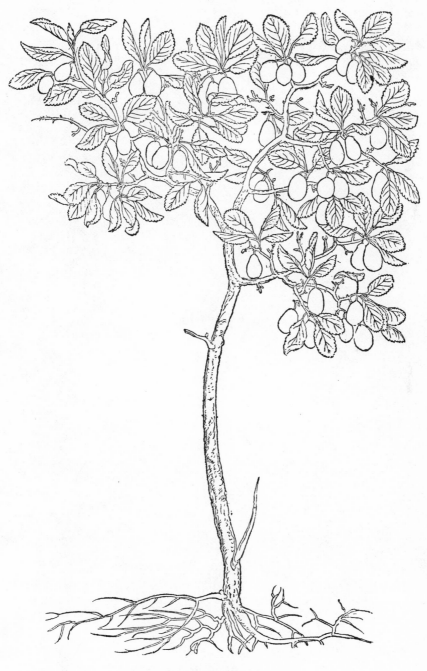

312.—PLUM.

PRUNUS DOMESTICA.   F. 403.

313.—SLOE. WILD PLUM. BLACKTHORN.
PRUNUS SPINOSA. M. (1565) 266.
Flowers white. Branches thorny, purplish, ending in a stout thorn. 15 ft. Mar.—Ap.
Fruit black, with bloom. A fine colour effect is produced when the green berries begin, partially, to turn to purple.

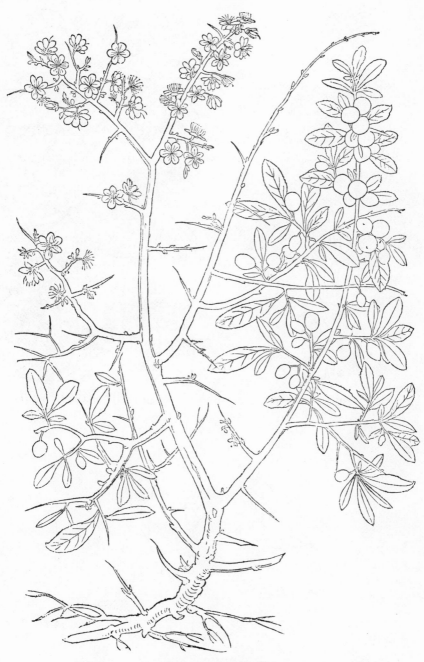

314.—SLOE, or BLACKTHORN.
Prunus spinosa. F. 404.

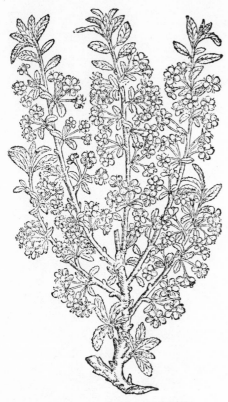

315.—THE SLOE IN FLOWER.
G. 1313.6.

316.—THE SLOE IN FRUIT.
G. 1313.5.

317.—COMMON LAUREL.
CERASUS LAUROCERASUS.
White. 20 ft. Fl., ⅜ in. Ap.—May.

318.—QUINCE.
CYDONIA VULGARIS. M. (1554) 132.2.
White. 20 ft. Fl., 1¼ in. May—June.

319.—QUINCE.
CYDONIA VULGARIS. G. 1264.
Fruit, orange-yellow, large. Nov.

320.—APPLE.
PYRUS MALUS. M. (1554) 132.1.
White, streaked with deep rose-colour. 20 ft. Fl., ¾ in. Ap.—May.

321.—APPLE.
PYRUS MALUS. L. ii. 165.1.

322.—APPLE-TREE.

PYRUS MALUS. C. S.

The Apple-tree makes a rounder head than the Pear-tree, and is not so erect. The flowers are in umbels, whereas those of the Pear are in corymbs. Nevertheless the corymbose formation is often present, but less straggling than in the Pear. The leaf is more oblong, and is bluntly serrated. The stipules are long and tapering, and produce a ragged effect. The leaf-stalk is shorter, wider, and more channelled than in the Pear.

QUINCE.—"A low tree with a crooked stem and tortuous, rambling branches." The fruit is ripe in November, of an orange-yellow colour, either of an Apple or Pear form, and large.

The JAPONICA is a shrub, 4 ft. high, from Japan. It is Cydonia japonica. Its petals are of the richest scarlet.

323.—PEAR.

PYRUS COMMUNIS. M. (1565) 251.

Flowers, white, ¾ in. The Wild Pear has thorns, and its leaves are serrated. Both these peculiarities are absent in the cultivated kinds. Flowers in a corymb. Stipules soon falling. The leaves and leaf-stalks are longer than in the Apple. The tree makes a much more pointed head than the Apple-tree, and some kinds reach a height considerably above 20 ft., which is the height of the Wild Pear. Ap.—May.

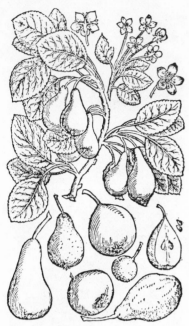

324.—PEAR.
PYRUS COMMUNIS. C. E. 152.

325.—"PEARE ROYALL."
G. 1268.3.

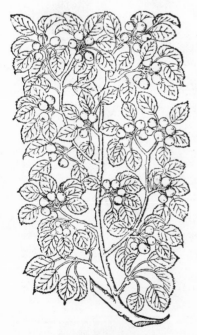

326.—WILD PEAR.
PYRUS COMMUNIS. G. 1271.3.

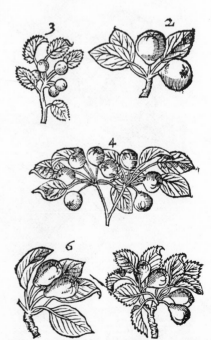

327.—PEARS OF DIFFERENT KINDS.
J. 1458.

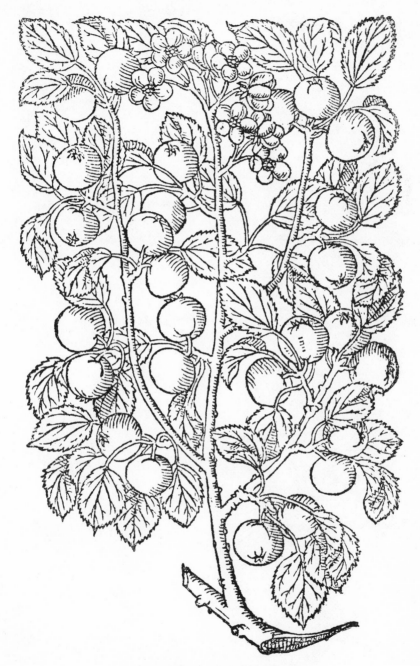

328.—"CHOKING LEAN CRAB-TREE."

PYRUS MALUS. G. 1277.4.

This figure is enlarged.

329.—"KING OF APPLES."
G. 1274.3.

330.—"KATHARINE PEAR-TREE."
G. 1269.7.

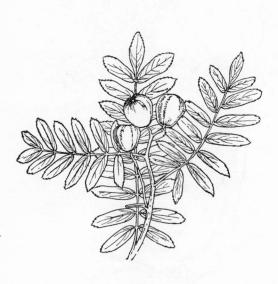

331.—SERVICE PEAR, OR SORBUS.
PYRUS DOMESTICA.
White. 30 ft. Fl., ⅝ in. Germany. May.

332.—WILD SERVICE.
PYRUS (OR SORBUS) TORMINALIS.
Yellowish-white. May.

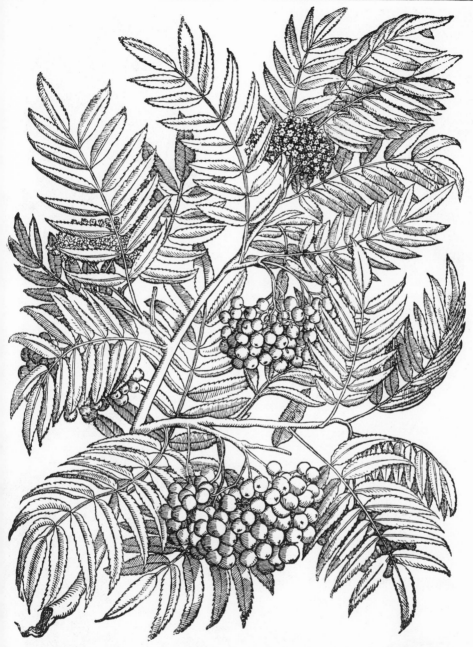

333.—MOUNTAIN ASH. ROWAN-TREE. QUICKEN-TREE.
Pyrus Aucuparia. M. (1565) 262.

The flowers of the colour of Meadow-sweet, and very fragrant, $\frac{1}{2}$ in. They are
here ill shown, and are better rendered in the next figure. This is one of the Beam-
trees—the timber was used for beams. 30 ft. May—June. Berries bright red.

334.—MOUNTAIN ASH.
PYRUS AUCUPARIA. D. 99.1.

335.—PYRUS SANGUISORBIFOLIA. D. 203.
(So identified by Stokes.)

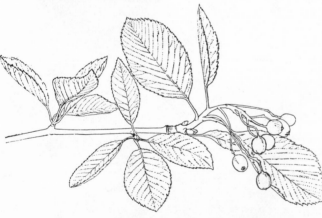

336.—WHITE BEAM. PYRUS (OR SORBUS) ARIA.

337.—CUMBERLAND HAWTHORN. PYRUS (OR SORBUS)
INTERMEDIA.

338.—COTONEASTER.
COTONEASTER VULGARIS. D. 198.
White, pink. 4 ft. Fr. Red. Ap.—M

339.—MEDLAR.

MESPILUS GERMANICA.   M. (1565) 253.

Flowers, large white.   12 ft.   Rare in England.   May—July.

340.—HAWTHORN. WHITE THORN. MAY.

CRATÆGUS OXYACANTHA. M. (1565) 163.

Flowers white or deep pink, in corymbs. Anthers pink, becoming brown. The pink flowers are rather paler behind. The flowers are here ill shown. It is evident that in this and other cases the drawing has been made when the plant was in fruit. 20 ft. ; but most common as a hedge bush. May—July.

341.—DETAILS OF HAWTHORN.
CRATÆGUS OXYACANTHA.

ROSES

Roses have for so long been objects of cultivation that the origin of the different kinds, if traceable, is not a matter of much interest. To the casual plant-lover roses are of three kinds—the Single Roses, the ordinary Double Roses, and the Rambling and Monthly Roses with their bunches of flowers. The wild English Roses are at least five—

R. spinosissima—Pimpernel, Burnet, or Scotch Rose. R. villosa—Apple or Downy Rose. R. canina—Dog-Rose or Hep Tree. R. arvensis—Corn, Field, or White Dog-Rose. R. rubiginosa—Sweet Briar.

From the first and the last many garden varieties have been produced. The cultivated roses have, however, chiefly come from certain exotics—the Damask Rose, R. damascena, a pink rose from the Levant ; the Gallican Rose, R. gallica, a red rose of South Europe ; the Yellow Rose, R. lutea, from Austria ; and the Provence Rose or Hundred-leafed Rose, R. centifolia. All but the last are single roses.

While the varieties of the rose are exceedingly numerous, we need only direct our attention to the normal characteristics of the genus, which are fairly constant. The branches or twigs are round, smooth, green, armed with stout, curved, reddish prickles, curved downwards, or with stiff prickly hairs. The leaves are pinnate, of usually 5 or 7 sessile serrated leaflets, of which those towards the base are smaller and directed backwards. It is a common error to draw the leaflets at right-angles to the mid-rib.

On either side of the leaf-stalk, perhaps occupying 1 in. of its length, and also slightly clasping the stem, are the stipules, which are membranous, and often toothed. They are a conspicuous and important part of the leaf. It may indeed almost be said that they are the most important part of the leaf, for in some cases the leaves get smaller and smaller up to the end of the twig against the flower, and do so by having fewer leaflets till, up against the flower, they will have but one leaflet, and finally none at all, but only the 2 stipules making a membranous leaf.

The calyx is pitcher-shaped, having a swollen ovoid part, which becomes the fruit, below, and 5 spreading points above which originally enclose the flower in the bud. These 5 points or segments sometimes have lobes down their edges—notably in Dog-Rose. The lobes are not equal upon all the segments—2 segments are usually without lobes, and 1 has them down one side only.

The stamens are numerous, much shorter than the petals—the anthers yellow. Styles numerous.

The petals are normally 5, but often numerous. They are roundish, or perhaps narrower, concaved, notched, and connected to the calyx by narrow claws. Two petals have both edges in front of their neighbours, and 2 have both edges behind.

In the Wild or Dog-Rose the uppermost leaves or bracts are opposite, and consist largely of the stipules.

342.—WILD ROSE.　　　　　343.—ROSE.

Z. 269.2.　　　　　Z. 269.1.

(Some further illustrations of Roses are given in the Appendix, Figs. 28 and 52.)

344.—DOUBLE WILD ROSE.
L. ii. 210.2.

345.—MUSK ROSE.
L. ii. 208.2.

346.—PIMPERNEL ROSE.
G. 1088.3.

347.—ROSE.
J. 1261.2.

348.—ROSE.
ROSA. M. (1565) 185.

349.—"DOUBLE CINNAMON ROSE."
J. 1268.7.

350.—ROSE.
C. S.

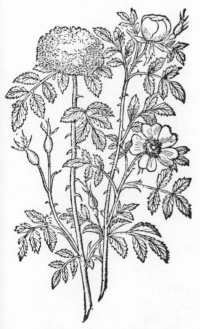

351.—BRIAR ROSE. DOG ROSE.
ROSA CANINA. L. ii. 210.1.

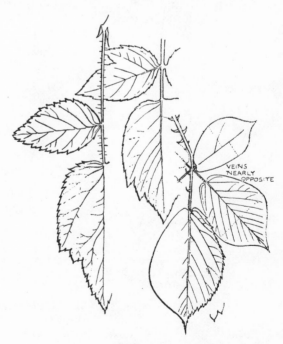

352.—BLACKBERRY LEAVES.

353.—BLACKBERRY BRAMBLE.

RUBUS FRUTICOSUS. F. 152.

Flowers white or pink ; ½ to ¾ in. July—Aug.

A scrambling shrub throwing its branches widely about in arching curves—in fact this drawing might more appropriately be placed upon its side ; branches green or purple or crimson, ribbed, often taking oot. This figure is evidently meant to represent two species. Perhaps the larger is meant for Dewberry.

354.—STRAWBERRY.

FRAGARIA VESCA. F. 853.

Flowers white, ½ in.   Height 9 in. ; creeping by runners.   Fruit red or white.   May—Aug.

The double calyx, consisting of 5 small points alternate with the 5 larger segments, is a beautiful feature not well shown here.

The BARREN STRAWBERRY.—Potentilla Fragiastrum—is very similar in general appearance except as regards the fruit.   Its leaves are very hairy on both sides.

355.—STRAWBERRY.
Fragaria vesca. G. 844.1.
White. 9 in. Fl., ½ in. May—Aug.

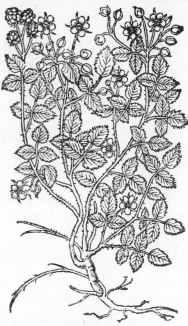

356.—RASPBERRY.
Rubus Idæus. G. 1089.2.

357.—COMMON CINQUEFOIL.
Potentilla reptans. G. 836.1.
Yellow. 4 in. June—Sep.

358.—"PURPLE CINQUEFOIL."
G. 836.3.

359.—COMMON, or CREEPING, CINQUEFOIL.

POTENTILLA REPTANS.　F. 624.

Flowers yellow ; ¾ in.　June—Sep.

360.—WHITE CINQUEFOIL.
POTENTILLA ALBA.　D. 1265.1.
White.　6 in.　June—Sep.

361.—WHITE CINQUEFOIL.
POTENTILLA ALBA.　G. 839.9.
White.　6 in.　June—Sep.

362.—SILVERY POTENTILLA.
POTENTILLA ARGENTEA.　G. 838.7.
Yellow.　8 in.　June—Sep.

363.—TORMENTIL.
POTENTILLA TORMENTILLA.　G. 840.
Yellow.　9 in.　June—July.

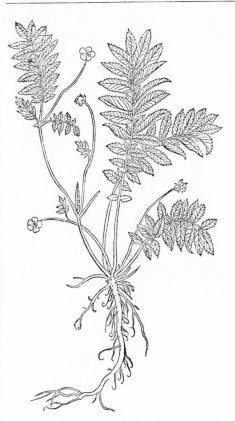

364.—SILVERWEED. GOOSEGRASS.
POTENTILLA ANSERINA. F. 619.
Yellow. Fl., 1 in. Lf., 5 in. June—July.

365.—SPRING POTENTILLA.
POTENTILLA VERNA. G. 838.8.
Yellow. 4 in. Fl., $\frac{5}{8}$ in. Ap.—May.

366.—MARSH CINQUEFOIL.
COMARUM PALUSTRE. G. 836.4.
Dingy purple. 1 ft. Fl., $\frac{3}{4}$ in. July—Aug.

The POTENTILLEÆ, or Strawberry group, include
Dryas, Geum, Potentilla, Fragaria, Rubus, Comarum
and Agrimonia. Comarum is now placed in Potentilla,
and Comarum palustre becomes Potentilla palustris.
Agrimonia differs from the others in having its seeds,
which are 2, enclosed within its hardened calyx.
It has been classed with Burnet and Lady's Mantle
as one of the Sanguisorbaceæ on account of this formation of the fruit.

The large points of the flower of Marsh Cinquefoil,
Fig. 366, erroneously given as six in number, are the
calyx.

367.—AGRIMONY.
AGRIMONIA EUPATORIA. F. 244.
Yellow. 2 ft. June—July. Britain.

The spikes should be quite upright, and rise proportionately much higher above the leaves. The drawing was, no doubt, made from a withered specimen.

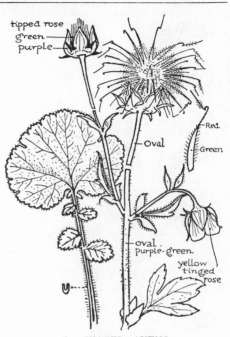

368.—HERB BENNET. AVENS.
GEUM URBANUM.
Yellow.　2 ft.　Fl., ⅝ in.　June—Aug.

369.—WATER AVENS.
GEUM RIVALE.
Orange.　Calyx purple.　1 ft.　Fl., ½ in.　June—July.

370.—WHITE DRYAS.
DRYAS OCTOPETALA.　J. 659.6.
Petals, 8.　White.　6 in.　Fl., 1¼ in.　June—July.

371.—WHITE DRYAS.
G. 533.4.

372.—MEADOW-SWEET.
SPIRÆA ULMARIA. L.V.K.
Yellowish-white.　3 ft.　Fl., ⅜ in.　July—Aug.

Many handsome species of SPIRÆA are to be
seen in gardens and in green-houses.　Their
flowers are generally pink.　Their fruit consists
of several follicles, or leathery capsules of 1
valve, bursting lengthwise, and reminding one
of such fruit as that of Columbine, which is one
of the Ranunculaceæ.　Spiræidæ are shrubs,
rarely herbs.

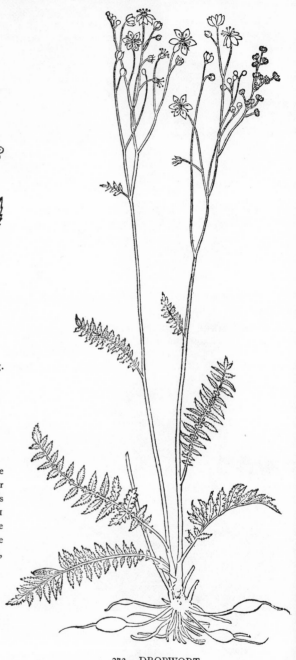

373.—DROPWORT.
SPIRÆA FILIPENDULA.　F. 562.
White or Pinkish.　1 ft.　Fl., ½ in.　July.

374.—SALAD BURNET.

POTERIUM SANGUISORBA. F. 789.

Height, 2 ft. Flowers crowded in a greenish globular head. Petals, 0. The upper flowers have crimson pistils, the lower have 30 to 40 drooping stamens, with purple filaments and yellow anthers. The plant is used as a salad. July—Aug.

COMMON BURNET—Sanguisorba officinalis—is a larger plant. It has an oval head of purplish flowers with 4 stamens only.

375.—LADY'S MANTLE.

ALCHEMILLA VULGARIS.

Green. Calyx 4-parted. Stamens, 4. Leaves folded fanwise, white beneath. 9 in.
May—Aug.

LYTHRACEÆ—LOOSE-STRIFE.

Herbs, often with a quadrangular stem, with simple willow-like leaves without stipules, and crumpled petals. The leaves are usually opposite. The calyx is tubular or pitcher-shaped, the limb divided into, say, 6 points. Adhering to the calyx are as many petals as the calyx has points, and as many or twice as many stamens. Style, filiform ; stigma, head-like. Capsule of one or many cells surrounded by the calyx.

The flower in the figure is inadequately rendered. See Fig. 54 in the Appendix.

Yellow Loose-Strife is one of the Primulaceæ.

376.—PURPLE LOOSE-STRIFE.

LYTHRUM SALICARIA. G. 386.2.

Purple. 4 ft. Fl., 1 in. July—Aug.

SAXIFRAGACEÆ—SAXIFRAGE.

Plants which often grow in patches. Leaves entire or divided, alternate, with or without stipules, often of a thickish character. The flowers are borne either singly or in panicles. The calyx is of 4 or 5 sepals more or less joined together at the base ; the corolla is of 5 petals (sometimes absent) inserted between the lobes of the calyx. Stamens 5–10. Ovary consisting of 2 carpels cohering below, diverging at the apex. Each apex bears a sessile stigma.

Lindley summed up the true Saxifrages thus—" Little herbaceous plants, usually with white flowers, cæspitose (growing in a little tuft) leaves, and glandular stems: some of the species have yellow flowers, others have red, but none blue." None are tropical.

GRASS OF PARNASSUS is now brought into this Order. HYDRANGEA is very nearly allied.

Nos. 380 & 382 resemble London Pride, which is Saxifraga umbrosa.

377.—WHITE MEADOW SAXIFRAGE.
SAXIFRAGA GRANULATA.   C. E. 719.

378.—WHITE MEADOW SAXIFRAGE.
SAXIFRAGA GRANULATA.
White. 1 ft. Fl., ⅝ in. May—June.

379.—YELLOW MARSH SAXIFRAGE.
SAXIFRAGA HIRCULUS.   J. 1284.8.
Yellow, with red dots. 9 in. Aug.

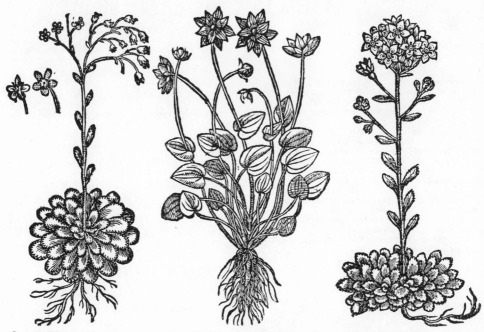

380.—SAXIFRAGE. Z. 1115.2.

381.—GRASS OF PARNASSUS.
PARNASSIA PALUSTRIS. J. 840.2.
White. 9 in. Fl., ¾ in. Aug.—Oct.
(For the flower see Appendix, Fig. 54.)

382.—SAXIFRAGE. J. 529.4.

CRASSULACEÆ—HOUSE-LEEKS.

Succulent herbs or shrubs. Leaves thick, entire or pinnatifid, without stipules. Flowers starry, usually in cymes, sessile, often arranged on one side of the stalk only. Sepals, petals, and carpels 3–20, stamens the same, or double the number, carpels tapering into stigmas. The plants are remarkable for their colour, which is sometimes blue-grey with pink flowers as in House-Leek; sometimes green with yellow flowers as in Stone-Crop, etc.

HOUSE-LEEK has 12 petals. Its leaves are in rose tufts, except upon the flowering stalks, which rise to 8 or 10 inches. WALL-PEPPER or STONE-CROP has 5 petals. ENGLISH STONE-CROP, Sedum anglicanum, is similar to it, but has white petals with red lines upon them, purple anthers, and the leaves and stems reddish. ORPINE has 5 petals. ROSE ROOT (MIDSUMMER MEN) has yellow flowers in a close terminal cluster. Its root, except when cultivated, has the fragrance of the rose. Its leaves are tipped with red—a peculiarity very common in this Order. Of Sedum there are several British species reaching to a foot in height.

COTYLEDON has round leaves (hence PENNYWORT) with somewhat lobed edges, and pendulous greenish flowers.

For Cobweb House-Leek, an Italian plant, see Fig. 14 in the Appendix.

384.—WHITE ORPINE.
SEDUM TELEPHIUM.   G. 416.2.

383.—PURPLE ORPINE.
SEDUM TELEPHIUM.   F. 801.
Reddish-purple.   18 in.   Aug.

385.—STONE-CROP.   WALL-PEPPER.
SEDUM ACRE.   G. 415.
Yellow.   3 in.   June—July.

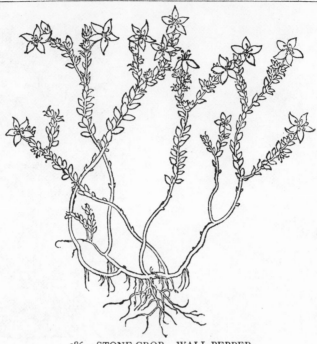

386.—STONE-CROP. WALL-PEPPER.
Sedum acre. F. 36.
Yellow. 3 in. June—July.

387.—HOUSE-LEEK.
Sempervivum tectorum. M. 490.1.
Pink. 8 in. Fl., ¾ in. July.

388.—ROSE-ROOT.
Sedum Rhodiola. F. 665.
Yellow. 8 in. June—July.

PASSIFLORACEÆ—PASSION-FLOWERS.

Herbs or shrubs, usually climbing by tendrils. Leaves alternate with foliaceous stipules. Flowers axillary or terminal, often with a 3-leafed involucre. Sepals 5, combined in a tube, the sides and throat of which are lined by a number of filamentous processes, "which are apparently metamorphosed petals," and form a "coronet." Petals 5, arising also from the calyx, outside of the processes. Stamens 5, monadelphous, surrounding the stalk of the ovary, the anthers turned outwards. The oval ovary bears 3 styles of curved form with dilated stigmas.

Passiflora incarnata, the ROSE-COLOURED PASSION-FLOWER, or MAY APPLE, is a twining plant reaching to 30 ft. in height, having pink, sweet-scented flowers, and bearing an orange-coloured fruit of the size of an apple. Native of America. Flowers in July and August.

The plant represented in Fig. 390 is styled Granadilla, which is the SWEET CALABASH, a West Indian dessert fruit—P. maliformis. But it appears more to resemble P. cærulea, the hardiest, and therefore the commonest, in English gardens. The flowers are blue outside, white and purple within.

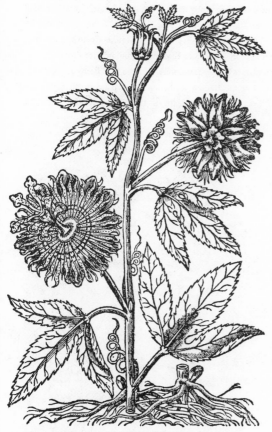
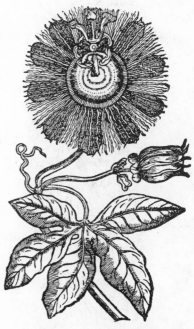

389.—ROSE-COLOURED PASSION-FLOWER.
PASSIFLORA INCARNATA. J. 1592.

390.—COMMON PASSION-FLOWER.
PASSIFLORA CÆRULEA? Z. 685.

[Here end the PERIGYNOUS CALYCIFLORÆ, with the flower around the ovary—practically below, or inferior to, it ; and here follow the EPIGNOUS CALYCIFLORÆ, which have the flower above or superior to the ovary—a condition well seen in the *Cucurbitaceæ*. Note that the *Rosaceæ*, in spite of the apparently inferior ovary in Rose itself, belong to the *Perigynæ*.]

CUCURBITACEÆ—MELONS.

Herbs with brittle stems, usually trailing over the ground, or climbing by means of tendrils, which occupy the position of stipules.   The whole plant is more or less succulent, the fruit being particularly so.   The stems are channelled, and they and the leaves are generally rough with hairs.

The calyx is insignificant, 5-parted, the corolla also 5-parted, sometimes of a crumpled texture, white, yellow or red, sometimes also small and herbaceous.

The flower is wholly above the ovary, which shows as a young fruit, more conspicuously beneath the flower than usual.   The fruit is afterwards crowned with the remains of the calyx.   Flowers unisexual.   Stamens 5, arranged in 1, 2 or 3 brotherhoods.   Anthers long and much bent about.

The whole Order is regarded as poisonous, or, at least, powerfully active, but some kinds, through cultivation, have become articles of food—the Cucumber, Cucumis sativus ; the Melon, Cucumis Melo ; the Water-Melon, Cucurbita Citrullis ; the Vegetable Marrow, Cucurbita ovifera succada ; the Red Gourd, Cucurbita maxima, and the White Gourd, Cucurbita Pepo.   The young shoots of the Common Bryony, Bryonia dioica, are used as a pot-herb, and in flavour resemble asparagus, though its roots are highly poisonous.   In the Middle Ages the roots were made to grow within moulds, till they took somewhat the human form, and were palmed off as Mandrakes. The Loofa, Loufah, or Luffa, which is used as a flesh-brush in the bath, is the fibrous inside of the fruit of Luffa ægyptiaca.

The leaves in all species are more or less palmate, with prominent veins.

391.—BALSAM APPLE.
Momordica Balsamina.   E.
Fruit warty, orange-red.   Syria and India.

392.—RED BERRIED BRYONY.
Bryonia dioica.   L. V. K. 311.

393.—RED BERRIED OR WHITE BRYONY.
BRYONIA DIOICA. F. 94.

The only British species. Scrambles over hedges. Leaves and tendrils opposite. Flowers in the axils of the leaves. Male and female flowers on different plants. Male, several together in a stalked raceme ; female, 2 or 3 together on short stalks—the former with 5 stamens and a campanulate corolla ; the latter with a 3-cleft style and a rotate corolla. Corollas yellowish-white, with green streaks. Berries green, changing to orange or red. Leaves, 2 in. Flowers, $\frac{1}{2}$ in. May—August.

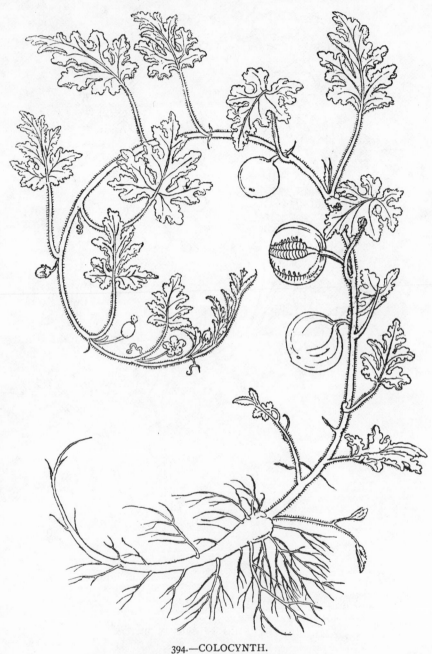

394.—COLOCYNTH.

CUCUMIS COLOCYNTHIS.   F. 372.

Flowers yellow.   Fruit the size and colour of an orange.

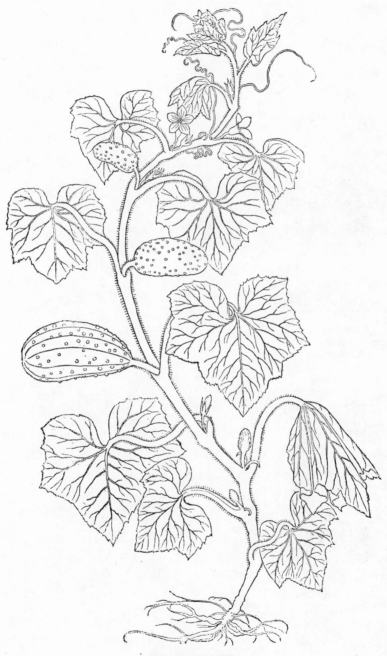

395.—CUCUMBER.
Cucumis sativus. F. 697.
Flowers yellow. 4 ft. E. Indies. July—Sep.

396.—PUMPKIN.
Cucurbita Pepo. F. 699.
Yellow. 16 ft. Levant. June—Aug.

397.—WATER-MELON.
Cucurbita Citrullus. F. 700.
Yellow. 6 ft. S. Europe. May—Sep.

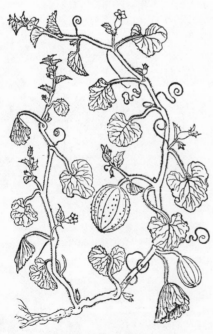

398.—MELON.
Cucumis Melo. F. 701.
Yellow. 4 ft. May—Sep.

399.—OBLONG GOURD.
Cucurbita Lagenaria oblonga. F. 370.
White. 10 ft. E. Indies. July—Sep.

400.—BOTTLE GOURD.
CUCURBITA LAGENARIA. F. 368.
Flowers white. Extent 10 ft. July—Sep.

' *Cucurbita*—a Latin word signifying a vessel. *C. lagenaria* has a fruit shaped like a bottle, [here the upper part is too wide for that description—but the fruit varies greatly] with a large roundish belly, and a neck very smooth ; when ripe, of a pale yellow colour. The rind becoming hard, and being dried, contains water. . . . It grows in all parts of Egypt and Arabia, wherever the mountains are covered with a rich soil."—*Loudon.*

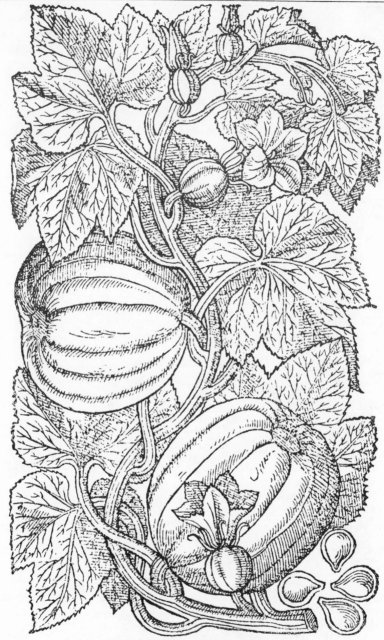

401.—"THE GREAT ROUND POMPION."
CUCURBITA PEPO. G. 773.2.

Loudon speaks of the plant (which is the Pumpkin) extending forty or fifty feet in a season, and covering an eighth of an acre. He speaks also of the fruit being, in England, filled with sliced apples, with sugar and spice, and fried and eaten with butter—the seeds having been removed through a hole cut for the purpose. This dish was called pumpkin pie. Parkinson (*Paradisus*, p. 526) referred to this country dish. So did Gerarde, in 1597—"but baked with apples in an oven . . . is food utterly unwholesome for such as live idly ; but unto robustious and rusticke people, nothing hurteth . . ." He also said—"The nourishment that commeth hereof is little, thin, moist and colde (bad, saith Galen), and that especially when it is not well digested : by reason whereof it maketh a man apt and readie to fall into the disease called the Cholerike passion, and of some the Felonie" (p. 775).

This figure is enlarged.

402.—POMPION, or PUMPKIN. VEGETABLE MARROW.
CUCURBITA PEPO. F. 698.

Flowers yellow, creamish. Native of the Levant. Cultivated in England in 1570.

The plant is sometimes trailed up wires against walls, and over doorways. Our common Vegetable Marrow is less clearly oval in form than this. It is called C. ovifera, C. ovifera succida, and C. esculenta. It is either a separate species, or a variety of C. Pepo.

CACTACEÆ—CACTUS.

Succulent plants with globular, flattened, or 3 or more angled stems, without leaves and bearing sessile handsome flowers, with the sepals and petals numerous and indistinguishable. Stamens numerous. Style, 1, with several stigmas. Fruit succulent, the best known kind being the Prickly Pear. Natives of tropical America. See Appendix, Fig. 54.

GROSSULARIACEÆ—CURRANTS AND GOOSEBERRIES.

Shrubs, some with thorns. Leaves alternate, lobed, toothed, and serrated—"often with a membranous edge to the base of the stalks." Flowers in axillary racemes, each with a bract at its base. Calyx, superior, 4- or 5-parted, regular, coloured. Petals, minute, inserted alternately with the stamens into the throat of the calyx. Style with a 2-lobed stigma.

For details see Appendix, Fig. 54.

The Flowering Currants are amongst the prettiest of shrubs. They are early in flower—March to May.

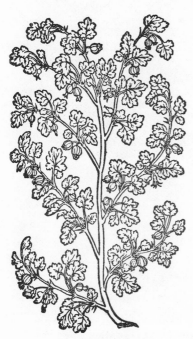

403.—GOOSEBERRIES.
Ribes Grossularia. Z. 264.

404.—CURRANTS.
Ribes rubrum. Z. 266.

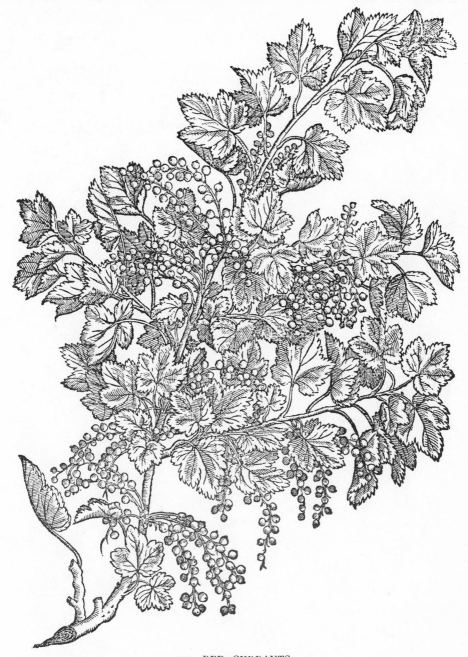

405.—RED CURRANTS.

RIBES RUBRUM. M. (1565) 168.

Flowers green. 4 ft. Ap.—May. In mountainous woods, especially about the banks of rivers,
in the North of England, and in Scotland.

The RED and the WHITE CURRANTS are varieties of the same species. BLACK CURRANT, Ribes nigrum,
grows in sandy swamps and thickets, about the banks of rivers. Its clusters are hairy, and are distinguished from
the other kinds by having a solitary flower at the base of each cluster.

227

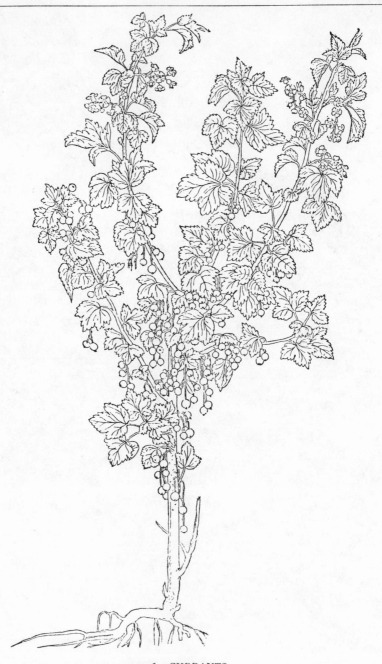

406.—CURRANTS.

RIBES RUBRUM.   F. 663.

Jacob Bobart, to whom the copy of Fuchsius, from which this figure is reproduced, belonged, wrote against it—"Bastard Currants," by which he may mean that he takes it to be one of the wilder kinds.

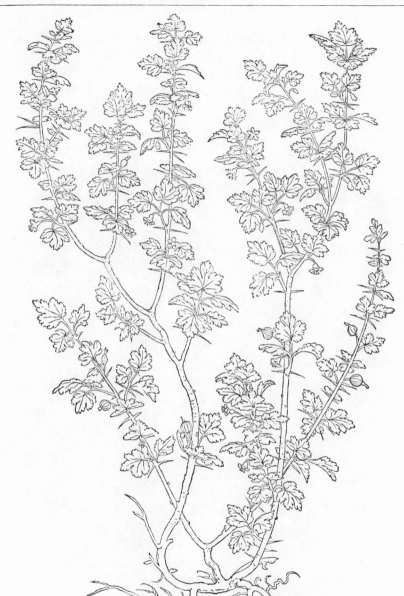

407.—GOOSEBERRIES. FEABERRIES.
RIBES GROSSULARIA. F. 187.

Flowers green. 4 ft. Ap.—May. Prickles, 1, 2, or 3 under each bud, but none on the intermediate spaces. Leaves smaller, rounder and smoother than in the Common Currant. Berries, green or yellowish, sometimes reddish, rough with scattered hairs, or smooth. Grows wild in hedges, thickets and waste ground, as well as cultivated in gardens.

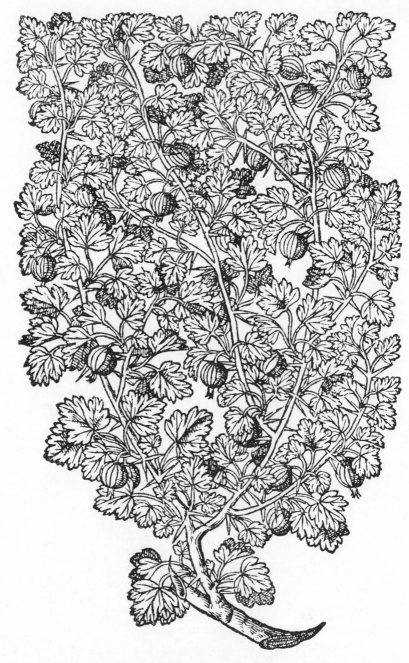

408.—GOOSEBERRIES.
RIBES GROSSULARIA. G. 1143.
This figure is enlarged.

PHILADELPHACEÆ—SYRINGA.

Shrubs with opposite leaves without stipules. Flowers with a superior calyx with 4 or more divisions, and a corolla of as many petals as the calyx has divisions. The SYRINGA or MOCK ORANGE has been cultivated as an ornamental shrub for three hundred years. Its native place is supposed to be Southern Europe. It is called Mock Orange from the resemblance of its flowers to those of the orange, both in appearance and odour. Another drawing is given in the Appendix, Fig. 53.

409.—SYRINGA. MOCK ORANGE.
PHILADELPHUS CORONARIUS. Z. 256.1.
White. 8 ft. May—June.

MYRTACEÆ—MYRTLE.

An Order including the MYRTLE, EUCALYPTUS, CLOVE-TREE (Caryophyllus aromaticus), and POMEGRANATE, which, however, is placed sometimes by itself in an Order of its own—the *Granatæ*.

Leaves, entire, opposite or alternate, usually with a vein close within the margin. Calyx 4- or 5-cleft. Petals 4 or 5. Stamens 8, 10, or many. Style and stigma simple. Flowers red, white, occasionally yellow, never blue, usually in the axils of the leaves. Fruit dry or fleshy.

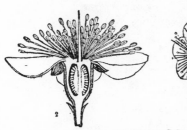

410.—FLOWER OF MYRTUS COMMUNIS.
L. V. K. 738.

411.—COMMON MYRTLE.
MYRTUS COMMUNIS. G. 1226.

412.—BOX-LEAFED MYRTLE.

MYRTUS COMMUNIS TARENTINA. M. (1565) 229.

Flowers white. 6 ft. S. Europe. July—Aug.

Berries green changing to black.

413.—BOX-LEAFED MYRTLE.
MYRTUS TARENTINA. C. E. 133.

414.—POMEGRANATE.
PUNICA GRANATUM. J. 1450.2.
Scarlet. 18 ft. Fl., 2 in. S. Europe. June—Sep.

ONAGRACEÆ—WILLOW-HERBS.

An Order embracing many beautiful plants—Fuchsia, Evening Primrose, and the Willow-Herbs. The number 4 prevails in the flowers. The ovary is beneath the flower, and in Epilobium forms apparently the stalk of it. The fruit is a berry or a capsule. The capsule is in general outline like that of the Crucifers, but opens from the top by 4 valves. The calyx is more or less tubular, 4-parted. In Fuchsia the four points curl back gracefully. In Evening Primrose it rapidly splits and turns back against the stalk. Petals 4. Stamens 4 or 8. Style filiform, with a capitate or a 4-lobed stigma. Leaves alternate, or opposite, sharply serrated, the stalks and leaf-veins often crimson.

CIRCÆA was held to be a plant used by the enchantress Circe in mixing her draughts, but there seems no ground for this beyond the similarity of the words, and of the leaf to that of Nightshade. Bentley says the plants of this Order are generally harmless. *Epilobium* is from the Greek epi, upon, and lobos, a pod—the flower upon a pod. *Fuchsia* is, of course, in honour of Leonhardus Fuchsius, many of whose excellent cuts are reproduced in this book, and who, in his *Historia Stirpium*, began the use of fixed botanical terms.

Epilobium hirsutum is called "Codlings and Cream," from the scent of its leaves resembling scalded apples.

For figures of Fuchsia and Willow-Herb see the Appendix, Figs. 53 and 54.

415.—ROSE-BAY WILLOW-HERB.
EPILOBIUM ANGUSTIFOLIUM.　D. 865.
Deep pink.　6 ft.　Fl., ¾ in.　July—Aug.
Petals unequal in size.

416.—GREAT HAIRY WILLOW-HERB.
EPILOBIUM HIRSUTUM.　D. 1059.3.
Pink.　5 ft.　Fl., ¾ in.　July.
Petals equal in size.

417.—ENCHANTER'S NIGHTSHADE.
CIRCÆA LUTETIANA.　J. 351.
Pale pink.　1 ft.　June—July.

418.—EVENING PRIMROSE.
ŒNOTHERA BIENNIS.　J. 475.
Yellow.　4 ft.　Fl., 3 in.　July—Sep.

234

CORNACEÆ—CORNEL-TREE.

Shrubs or trees, with opposite leaves, without stipules.　Flower superior.　Sepals, petals, and stamens 4, style 1.　Fruit a berry, crowned with the remains of the calyx.

419.—DOGBERRY.　CORNEL-TREE.
CORNUS SANGUINEA.　M. (1565) 260.
A shrub, 8 ft.　Flowers, white, in cymes.　Berries black.　Twigs red.　Woods, hedges.　June.

420.—CORNELIAN CHERRY.
CORNUS MASCULA.　M. (1565) 259.

A tree, 15 ft.　Flowers yellow.　Fruit cornelian-colour.　Austria.　Feb.—Ap.

The wood is hard and horny.　Its little clusters of yellow starry flowers (the sepals, petals, and stamens are 4) "stud its naked branches and are amongst the earliest heralds of spring."　The berries, which are like little plums, were formerly used for tarts.

UMBELLIFERÆ—HEMLOCK, PARSLEY, Etc.

No Order is so readily recognized as this, nor are other plants easily confused with it.   The smaller species of Thalictrum are perhaps the nearest to them in general outward appearance, partly owing to the leaves being twice-pinnate and of a triangular mass-shape, partly also to the arrangement of the carpels in Thalictrum suggesting at first sight an umbel-like construction.   But some of the Umbellifers do not immediately reveal their connection with their Order.   This is owing to their involucres being of unusual size and character, and to their flowers not being perched upon long pedicels, but clustered together in heads—their umbellate character is thus lost.   Of such plants, Common Sanicle, Astrantia, and Sea-Holly (Eryngium) are noteworthy.   COMMON SANICLE has leaves which suggest Lady's Mantle.   ASTRANTIA has its involucre of many regular white or rose-coloured leaves, which form a cup, and look like a flower—the small flowers being within.   In Eryngium also the involucre is very large, and the flowers are so close together that the plant looks more like a Composite, especially as it has a thistle-like prickliness.

As a rule, however, the flowers are in double umbels.   There is a *general*, or universal, umbel, which consists of a number of peduncles, each bearing a small umbel of pedicels, each of which supports a flower.   These smaller umbels are called *partial* umbels.

The partial umbels sometimes touch one another and make one broad mass ; sometimes, however, there is some space between them.   Frequently the flowers at the edge of the umbel have 1 or 2 of their petals larger, and make a gay border to it.   The stalks which compose the general umbel are not all of the same length ; those at the outside are longer, so that the whole mass of flowers is flat rather than globose.   But in this matter there is considerable difference.   In ANGELICA the umbels are globose.   When the CARROT goes to seed, the stalks of the umbels fold together, when, owing to the outer stalks being much longer than the inner, a form is produced like a bird's-nest (Fig. 433).

At the bases of the general and partial umbels there is usually an involucre, sometimes completely encircling the umbel, sometimes only partly present.   These involucres afford considerable indication of species, but only in a few cases are they conspicuous—notably in FOOL'S PARSLEY, Æthusa Cynapium, where every partial umbel has an involucre of 3 narrow long pointed leaves hanging down, all placed on the outer side of the umbel ; and in Carrot, Fig. 434.   The general involucre is often absent.

The flowers, though very small, possess great beauty.   They have a corolla of 5 petals, superior upon the ovary.   The petals are heart-shaped, and notched.   Often the notch is caused by the point being bent over on to the front.   As has been remarked, sometimes the petals at the margin of the umbel are larger than the others.   The stamens are 5, placed between the petals.   The styles are 2.

The flowers are generally white, sometimes the colour of Meadow-Sweet, often pale pink or pink, in a few cases yellow, and in Angelica archangelica green.

There is a common tendency to pinkiness in the stems, sheaths and flowers, but not in the leaves.   In HEMLOCK and GIANT HEMLOCK the stems are splashed with red-purple.

The fruit consists of 2 carpels, crowned by the 2 styles, and conjoined by their flat surfaces.   They are supported by a 2-branched axis.   The two branches pass up between the carpels.   When the fruit ripens the carpels separate, each hanging from the summit of one of the branches of the axis.   The carpels are more or less oval in general outline.   They vary much in shape, and are distinguished botanically by the ribs upon their rounded outer surfaces, and by certain reservoirs of oil which most of them contain.   Some, as Carrot, have rows of bristles or teeth.

The stems are furrowed, sometimes smooth and shiny, as in Hemlock, sometimes rough and bristly.   In some cases they are swollen below the joints.

The leaves, although sometimes simple, are also pinnate, but most frequently twice-pinnate.   They often make a triangular mass-shape, and may be said to be ternate.   The leaflets are generally toothed, and somewhat serrated ; and the divisions are regular, and in some plants suggest ferns.   At its base the leaf has often a baggy membranous sheath in which is enclosed the next leaf, curled up, and also a flower-head or two.   In Common Hogweed, or Cow-Parsnip, these sheaths are very noticeable, and are usually reddish or purplish in colour.

Some of these plants are esculent, others aromatic and stimulant, others poisonous, and others yield gum-resins.

The chief of the Esculent plants are the Parsnip, Carrot, Parsley, Celery, Chervil, Fennel and Samphire ; of the Aromatic—Dill, Caraway, Coriander, Fennel, Anise ; of the Poisonous—Hemlock, Water-Hemlock, Dropwort and Fool's Parsley.

The British species number 59, and are hard to distinguish.   Space does not permit their being even enumerated.   Among the commonest are WILD CHERVIL, COW-WEED or COW-PARSLEY, Chærophyllum sylvestre ; ROUGH CHERVIL, Chærophyllum temulum ; HOGWEED or COW-PARSNIP, Heracleum Sphondylium ; and HEMLOCK.   The selection of figures given will probably give a sufficient idea of the Order for the designer's purposes.

The form of the leaf varies from the finely segmented of FENNEL to the round shield-like one of MARSH PENNYWORT—Hydrocotyle vulgaris, a common little plant found in moist places.

421.—ASTRANTIA MAXIMA.
Flowers and inner side of involucre pink ; outer side white, veined green.   2 ft.   Caucasus.   June—July.

422.—WOOD SANICLE.
Sanicula Europæa. F. 671.
Florets sessile, white.   Lower leaves palmate, the lobes trifid, serrated.   Height, 1 ft.   Woods.
June—July.   The name is from *Sanare*—to cure, and the plant was esteemed a powerful vulnerary.

423.—GREAT BLACK MASTERWORT.
ASTRANTIA MAJOR.   F. 670.
2 ft.   Alps of Europe.   May—Sep.

Astrantia maxima is represented in Fig. 421. Its involucre is white with green veins ; the florets
are pink.   The involucre, as the flowers develop, falls back to some extent, when the star-like character
of the flower is better exhibited.

239

424.—PRICKLY SEA-PARSNIP.
ECHINOPHORA SPINOSA.   M. (1565) 490.
A sea-coast plant, here shown in fruit.   Not British, or very rare.   Height, 9 in.

   Stokes identified the same figure when given in Dalechampius as Crithmum maritimum, the ROCK
SAMPHIRE, a British sea-coast plant.   It is similar, but less spiny.   It grows from the crevices of rocks,
hanging down, usually in places difficult of access.

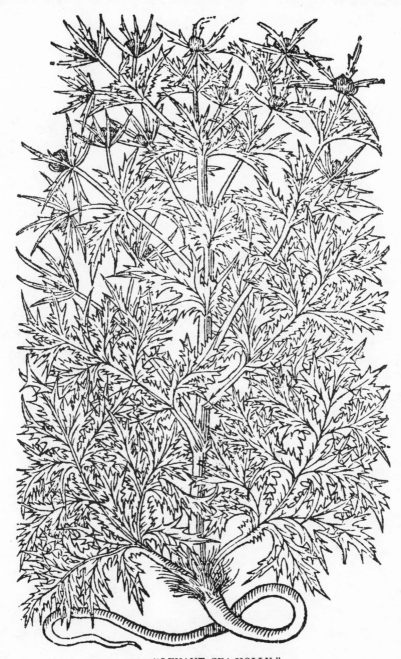

425.—"LEVANT SEA-HOLLY."

Eryngium campestre. G. 999.2.

This figure is enlarged in order that the leaves, which, though confused in effect, are admirably drawn, may be better seen. The figure indeed requires colouring. The plant appears to be the Common Eryngo. Its height is 2 ft., its flowers are purplish, and its leaves, unlike the Sea-Holly, which is glaucous, blue-grey, are green edged with yellow. It is a rare plant in England. July—Aug.

426.—COMMON ERYNGO.

Eryngium campestre. F. 296.

2 ft. Flowers purplish. July—Aug.

The Eryngos had much medicinal fame in the past. They were accounted aphrodisiac ; and good both for man and beast. Comfits used to be, and perhaps still are, made of the roots of this plant and of Sea-Holly. They are the "kissing comfits" of Falstaff. Gerarde gives instructions for making them, and says they are "exceeding good to be given unto old and aged people that are consumed and withered with age, etc."

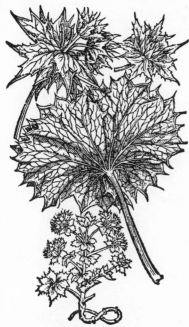

427.—SEA-HOLLY.
ERYNGIUM MARITIMUM.　C. E. 448.
Plant glaucous.　2 ft.　Fl., blue.　July.

428.—THOROUGH-WAX.　HARE'S-EAR.
BUPLEURUM ROTUNDIFOLIUM.　L. i. 396.1.
Yellow.　18 in.　June—July.

429.—"BROAD-LEAFED HARE'S-EAR."
BUPLEURUM.　D. 741.1.
A European species.　Fl., green.

430.—PARSLEY.
CARUM PETROSELINUM.　G. 861.1.
Yellow.　3 ft.　June—July.

431.—GREAT PIMPERNEL.　GREAT BURNET SAXIFRAGE.

PIMPERNELLA MAGNA.　F. 608.

3 ft.　Flowers white.　Woods.　Aug.—Sep.

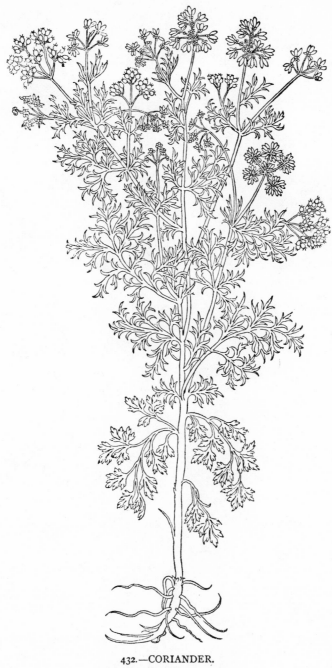

432.—CORIANDER.

CORIANDRUM SATIVUM. F. 345.

Inner flowers not fertile. Outer petals large. This plant is sometimes quoted as an instance of the uselessness of assigning a particular shape of leaf to a plant, as the upper leaves differ from the lower.

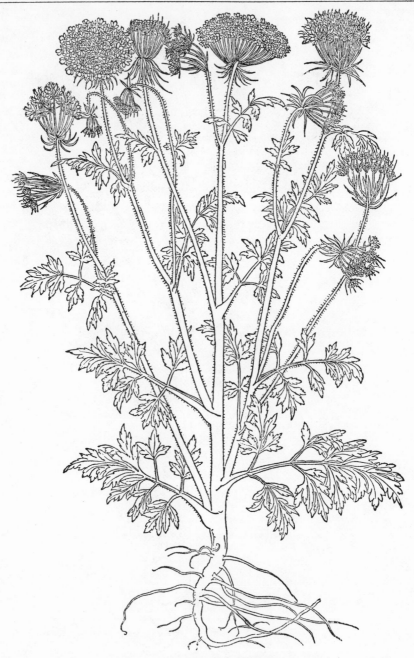

433.—CARROT.　BIRD'S-NEST.

Daucus Carota.　F. 684.

The outer flowers have irregular petals.　There are involucres to both general and partial umbels—the pinnate leaves of the general involucre are not well shown here, nor is the bird's-nest formed by the umbel when in fruit. Middle flowers not producing fruit.　The flower head, white or pale pink, sometimes with a deep pink spot in the middle of the whole umbel.　Fruit rough with bristles, and reddish-purple in colour, producing a handsome effect.

434.—WILD CARROT.
DAUCUS CAROTA.
Pinkish-white. 2 ft. July.

435.—ALEXANDERS.
SMYRNIUM OLUSATRUM. G. 866.
Yellow. 4 ft. May—June.

436.—VENUS'S COMB. SHEPHERD'S NEEDLE.
SCANDIX PECTEN-VENERIS. C. E. 304.
White. 1 ft. June—Aug.

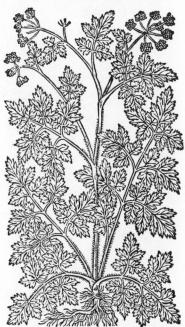

437.—ROUGH CHERVIL.
CHÆROPHYLLUM TEMULUM. J. 1038.2.
White. Stem spotted. 3 ft. July.

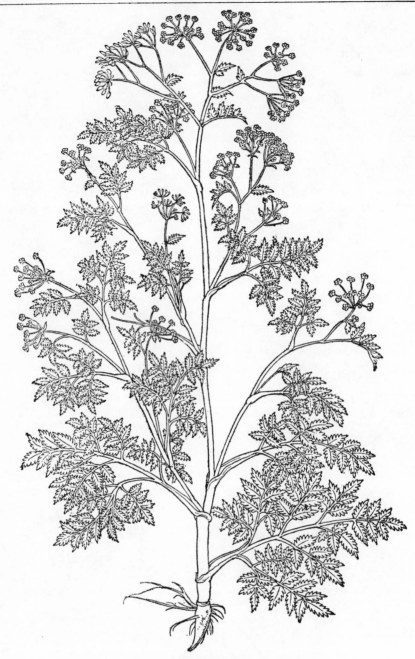

438.—HEMLOCK.
CONIUM MACULATUM. · F. 406.

3 ft. Stem smooth, spotted with purple. General umbel of 3 to 7 broad reflexed leaves, the partial umbel of 3 or 4, on one side of the umbel ; both very short. Leaves dark green, shining. Poisonous. The attachment of the umbels is, in places, here incorrect. Flowers white, all fertile, the outer ones irregular. June—July.

439.—PARSNIP.
PASTINACA SATIVA.　F. 751.

The flowers are yellow ; the partial umbels not touching one another.　The fruit is oblong, flatted, and surrounded with a membrane.　Height, 3-4 ft.　July.

This drawing, while giving a most admirable rendering of foliage, has the flowers but slightly expressed.

440.—COMMON MASTERWORT.
IMPERATORIA OSTRUTHIUM. F. 763.
Flowers pink. 2 ft. May—June.

This plant was reputed to be of great virtue, hence receiving the names Masterwort and Imperatoria.

250

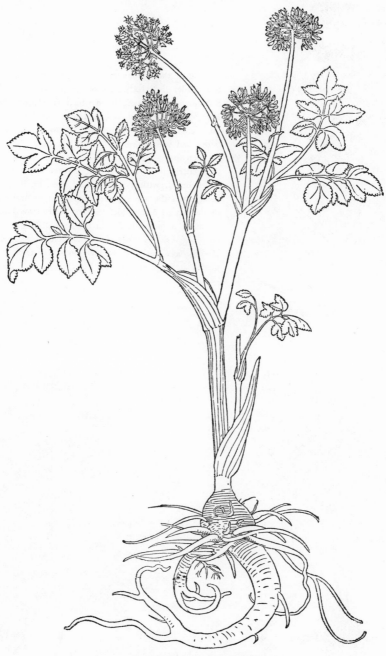

441.—GARDEN ANGELICA.

ANGELICA SATIVA. F. 124.

Umbels globose ; all the flowers regular and fertile, corollas greenish. Height 4-6 ft.
June—Sep. Not native in Britain. This is a larger plant than WILD ANGELICA—
Angelica sylvestris, which has narrower leaflets and corollas, and stems tinged with pink.
Its umbel is roundish, not globose. Height 2-4 ft. Flowers pinkish. July.

251

442.—GIANT HEMLOCK.
D. 790.1.

One of the commonest and most conspicuous plants of this Order is HOGWEED or COW-PARSNIP, Heracleum Sphondylium, for a figure of which see Appendix, Fig. 54. It is often called Hemlock, and bears some resemblance to the Giant Hemlock, which is well known as an ornament in parks, and which rises to 10 or 12 ft. high. The similarity between the two plants is in the form of the leaves and in the general bearing of them. Hogweed often ends in a large umbel, with 3 or 4 smaller ones, on long stalks, all rising from the same place around the central stem. But to the designer the chief interest in Hogweed and Giant Hemlock is in the leaf. In the leaf of Hogweed a cross-form, suggestive of the cross-crosslet of heraldry, is observable. If Fig. 442 really represents Giant Hemlock it does so inadequately. See Appendix Fig. 53.

443.—WATER-PARSNIP.
SIUM LATIFOLIUM. D. 1092.3.
White. 4–5 ft. Ditches. July—Aug.

444.—"HERCULES WOUNDWORT."
J. 1003.2.

ARALIACEÆ—IVY.

Trees, shrubs, and herbaceous plants closely connected botanically with umbellifers, from which they chiefly differ in having an ovary with more than 2 cells. They, moreover, "never possess to any degree the poisonous properties" of those plants.

The leaves are alternate, without stipules. The petals 2, 4, 5 or 10, or wanting. Stamens as many or twice as many as the petals, and alternate with them. Flowers usually in umbels or in heads.

Ivy (Hedera Helix) has a small calyx with 5 teeth, 5 petals, greenish-white, bent inwards at the points, 5 stamens.

"When it trails on the ground its branches are small and weak, and its leaves have 3 lobes; in which state it does not produce fruit, and has been called Barren or Creeping Ivy. When it climbs up walls or trees it grows much stronger, and the leaves change to egg-shaped—at least when arrived to its full growth, in which state it is called Climbing or Berried Ivy."

Bacchus is crowned with Ivy—it being reputed to prevent intoxication.

Ivy flowers in October and November; the berries are formed by February and ripe by April.

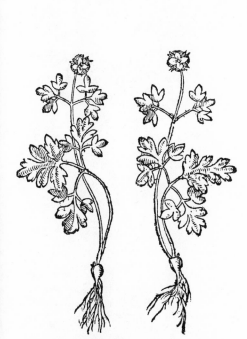

445.—TUBEROUS MOSCHATEL.
ADOXA MOSCHATELLINA. G. 933.10.
Green. 3 in. Berries reddish. Ap.—May.

446.—IVY.
Brunf. ii. 5.

447.—IVY.
HEDERA HELIX.　G. 708.1.

448.—IVY.
HEDERA HELIX.　G. 708.2.

[Here end the CALYCIFLORÆ, which have the stamens and petals upon the calyx ; and here begin the COROLLIFLORÆ, which have the stamens upon the corolla, which moreover is *monopetalous*. The Corollifloræ include practically all the monopetalous plants, but some few are among the Calycifloræ—the Cucurbitaceæ.]

CAPRIFOLIACEÆ—HONEYSUCKLE.

Shrubs with opposite leaves, without stipules, and flowers usually in terminal or axillary clusters, cymes, or whorls. Calyx, 4- or 5-cleft, usually with 2 or more bracts at the base. Corolla monopetalous, regular or irregular, sometimes regularly 4- or 5-cleft. Stamens as many as the lobes of the corolla. Style 1 ; stigmas 3 or 5. Fruit dry, fleshy, or succulent, crowned with the remains of the calyx.

The best-known plants are the Honeysuckle and the Elder. The Elder has its flowers in cymes with 5 branches, producing a flat head, not a roundish one as erroneously given in Fig. 449. The leaves are pinnate—the end leaflet slightly hidden by the first pair. Another tree is the MARSH ELDER, Viburnum Opulus, the GUELDER ROSE, which has its outer flowers large and barren. In its cultivated state all its flowers are large and barren, and it is then called the SNOWBALL TREE. Both kinds are shown in Fig. 451. The WAYFARING TREE, Viburnum Lantana, has pliant mealy twigs, and broad simple leaves. It reaches 10 ft., has white flowers, and is seen to be related to the LAURUSTINUS. WEIGELIA and SNOWBERRY are garden shrubs belonging to this Order. The former has clusters of large red and pink flowers ; the latter has small pink flowers, followed by large oval white berries. Of Honeysuckles there are several kinds differing considerably in the form of their flowers from the well-known species. Some of them are British.

For a drawing of the Honeysuckle, see the Appendix, Fig. 54.

449.—ELDER.
Sambucus nigra. M. 547.1.
White. 15 ft. May—July.

451.—SNOWBALL-TREE (*above*).
MARSH-ELDER (*below*).
GUELDER ROSE. Viburnum Opulus.
White. 10–15 ft. June. Z. 132.1.

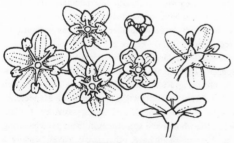

450.—FLOWERS OF ELDER.

Note.—A flower occurs in the vertex of each branching. These few flowers represent one of the final small
clusters into which the cyme branches. Contrary to the
case of umbels the middle flowers come out first, though
practically all are in bloom together.

452.—LAURUSTINUS.
Viburnum Tinus. D. 204.3.
White, buds pink. 4 ft. Jan.—April.

453.—ITALIAN HONEYSUCKLE.
Lonicera Caprifolium.   M. 440.
White or purple.   May—June.

454.—WOODBINE. HONEYSUCKLE.
Lonicera Periclymenum.   Z. 284. 1.
Pale yellow, red outside.   Berries red.
June—July.

RUBIACEÆ—MADDERWORTS.

A very extensive Order, " of such extent that it embraces a very large proportion of the whole of phænogamous plants." Lindley divided the Order into Cinchonaceæ, and Galiaceæ or Stellatæ. The former are tropical, the latter of the temperate regions. The Cinchonaceæ have opposite, entire, leaves with interpetiolary stipules and round stems ; the Galiaceæ have whorled leaves, without stipules, and have square stems.

Calyx, superior, with 4, 5 or 6 lobes. Corolla, monopetalous with the same number of divisions as the calyx. Stamens equal in number to the lobes of the corolla and alternate with them. Styles 2.

The chief plant of the Cinchonaceæ is the COFFEE PLANT—Coffea arabica. Jesuit's Bark is the bark of certain species of Cinchona, and Ipecacuanha is derived from Cephäelis Ipecacuanha. Some of the plants yield dyes. In this quality the Galiaceæ excel them, the foremost example being Rubia tinctoria, from the root of which Madder is obtained. Common amongst our wild flowers is LADY'S BEDSTRAW. Of these Wild Madders, or Bedstraws, there are 18 British species. The leaves are 4, 6 or 8 in a whorl, and in CROSS-WORT, Galium cruciatum, are broad. The flowers are sometimes close down in the whorls, and sometimes in loose panicles rising from either side of the whorls. The little branched stems bearing them have other small leaves, usually 2 together.

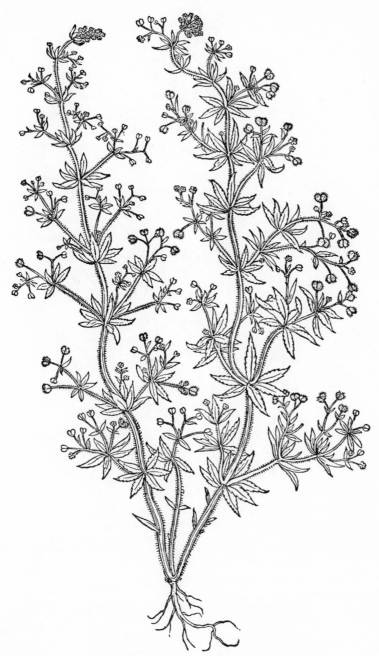

455.—GOOSEGRASS. CLEAVERS. HAYRIF.
GALIUM APARINE. F. 50.

Stem 4-cornered, the angles set with reflexed prickles, as also are the edges and keels of the leaves, which are 4 to 8 in a whorl. Fruit bristly—reddish. Flowers white. Hedges. May—June.

456.—YELLOW LADY'S BEDSTRAW.
GALIUM VERUM.　F. 196.

Leaves 8 in a whorl, smooth, their edges rolled back.　Roots produce a red dye.　Flowers yellow.　6 in. to 2 ft.　Roadsides.　July—Aug.

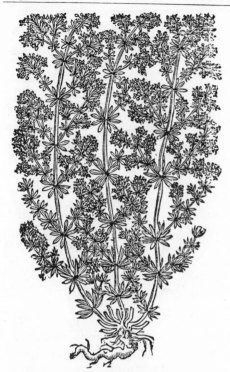

457.—WHITE LADY'S BEDSTRAW.
Galium Mollugo. G. 967.4.
White. 4 ft. Leaves, 8 together. July—Aug.

458.—MADDER.
Rubia Tinctoria. F. (1545).
Yellow. 4 ft. S. Europe. June.

459.—SWEET WOODRUFF.
Asperula odorata. J. 1124.1.
White. 1 ft. Woods. May.

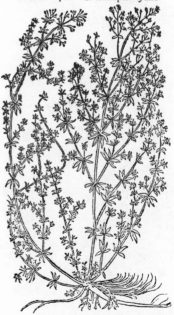

460.—WILD MADDER.
Galium Mollugo. F. (1545).
White. 4 ft. July—Aug.

VALERIANACEÆ—VALERIANS.

Herbs with opposite leaves without stipules, and usually either strong-scented or aromatic. Flowers with a superior calyx either with a membranous limb or one resembling feathery pappus. Corolla monopetalous, tubular, with 3 to 6 lobes, regular or irregular, often spurred at the base. Stamens 1 to 5. Style 1. Stigmas 1 to 3.

Valerians are nearly related to Teazels and Composites, but they do not so readily form a capitate inflorescence.

461.—RED SPUR VALERIAN.
CENTRANTHUS RUBER. D. 1187.2.

462.—GARDEN VALERIAN.
VALERIANA PHU. G. 917.1.
White. 3 ft. Germany. May—July.

463.—GREAT WILD VALERIAN.
VALERIANA OFFICINALIS. M. 28.2.
Pink. 4 ft. June.

464.—GREAT WILD VALERIAN.
VALERIANA OFFICINALIS.
(See Fig. 463 on the last page.)

465.—CORN SALAD. LAMB'S
LETTUCE.
VALERIANELLA OLITORIA.   G. 242.1.
Blue.   6 in.   Ap.—June.

DIPSACACEÆ—TEAZELS.

Plants differing from Composites in having opposite leaves, 4 instead of 5 stamens, with their anthers free instead of syngenesious, an involucel surrounding the base of each floret, and a simple stigma.   Nevertheless, it must be noted that some of the composites, as Zinnia, have opposite leaves.

The leaves are without stipules, and sometimes are confluent, and form a water-bowl around the stem.

The flowers are collected together upon a common receptacle, which is surrounded by a many-leafed involucre, of which some of the leaves are sometimes long and spiny.   The receptacle is covered with harsh bracts, with prickly points.   These bracts it is which give the Fuller's Teazel its usefulness.   Each flower consists of an ovary surmounted by a membranous or pappose calyx, and surrounded by a membranous involucel.   The corolla is inserted in the calyx.   It is tubular, 4- or 5-cleft, usually 4-cleft.   In Scabious the outermost corollas are large, and give the flower somewhat the appearance of Corn-flower.

FULLER'S TEAZEL is not a British plant, though sometimes found growing wild.   The WILD TEAZEL, Dipsacus sylvestris, has some of the spiny bracts of the involucre long and gracefully enclosing the head of flowers. The bracts of the flower-head itself are straight, and not hooked as in Fuller's Teazel.

FIELD SCABIOUS, Scabiousa arvensis, is similar to the small scabious, but is 2 to 3 ft. high, and its marginal corollas are more radiant.   July.

In the garden are many handsome and tall species, yellow, white, red, blue and purple.   SWEET SCABIOUS, Scabiosa atropurpurea, has a musky scent.

261

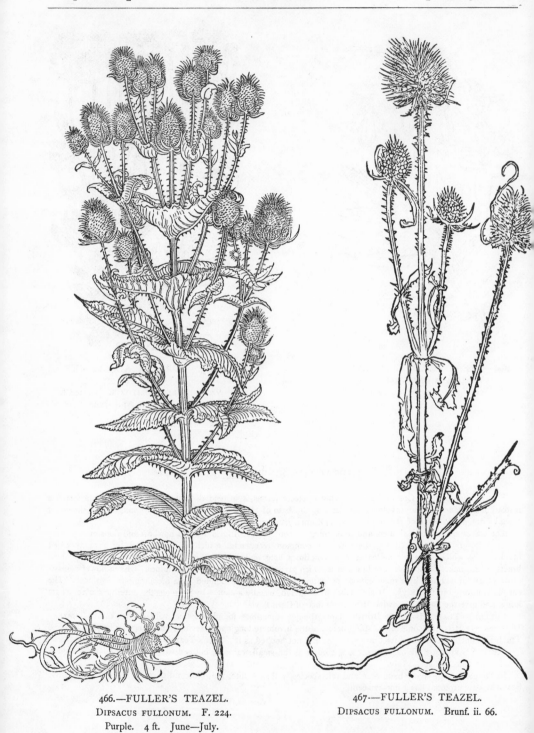

466.—FULLER'S TEAZEL.
DIPSACUS FULLONUM. F. 224.
Purple. 4 ft. June—July.

467.—FULLER'S TEAZEL.
DIPSACUS FULLONUM. Brunf. ii. 66.

468.—FULLER'S TEAZEL.
Z. 769.

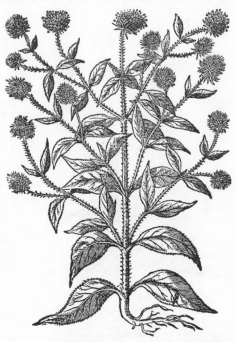

469.—SHEPHERD'S ROD. SMALL TEAZEL.
Dipsacus pilosus. M. (1565) 662.
White. 4 ft. Aug.—Sep.

470.—" RED SCABIOUS OF AUSTRICH."
G. 583.6.

471.—SMALL SCABIOUS.
Scabiosa columbaria. G. 582.1.
Mauve. 18 in. Fl., 1 in. June—Aug.

472.—DEVIL'S BIT SCABIOUS [Dipsacaceæ].

Scabiosa succisa. C. E. 397.

Blue. 1 ft. Fl., 1 in. Aug.—Oct.

COMPOSITÆ—DAISY AND DANDELION.

These plants are not easily confused with others. Anemones may be mistaken for Daisies, but they are readily seen not to be, upon examination, as they have no "calyx" behind what appear to be their petals, and their central construction is quite different.

Composites have tiny flowers, or florets, aggregated in close heads, upon a "receptacle" which is usually domed, and is often hairy, scaly, or chaffy. This receptacle is surrounded and supported by an involucre consisting of one or more rows of leaves or bracts. These bracts are sometimes long, as in Goat's-Beard, Nodding-Bidens, and some Asters; and sometimes they are white, or coloured, and horny, and have the appearance of being petals—as in the Everlasting Flowers.

The florets have at their bases a single bract or scale, called a *palea*, as particularly in Sunflower. But these paleæ are often absent. In the Teazels, in the previous Order, they are remarkable for their sharpness and hardness.

The floret consists of an ovary surmounted by a calyx, which is indistinguishable from it. The calyx is notable for its limb, which is membranous, divided into bristles or feathers—called *pappus*. Sometimes this limb is wanting.

The corolla is tubular, ending in 5 points, but sometimes it is slit down the side and becomes a strap, or "ligule." The ligule is sometimes quite strap-like with parallel edges, and ends in 5 little teeth, but sometimes its borders are rounded, and the end has perhaps only a notch. The ligule has several veins or channels upon it, normally 5, but sometimes fewer. Perhaps a ligule may best be regarded as having two channels and three points at the end—the middle one much smaller than the others. In some plants, as Sunflower, the ligule comes to a single point; in others, as Coreopsis, there are three points—the middle one the largest. In colour the ligule is sometimes different on its two sides, generally when this is the case it is paler and greener, or else darker, redder, browner, or purpler behind. This change of colour renders some plants, as some Chrysanthemums and Zinnias, particularly beautiful.

When so many flowers are massed together in so small a space, it is not unusual for the outer ones to develop for show. In this Order this is particularly the case, and there could be no greater distinction than that between the outer and inner florets of, say, the Marguerite Daisy, or the Sunflower. The florets are called disc-florets, and ray-florets, according as they occupy the middle or edge of the receptacle. The ray-florets

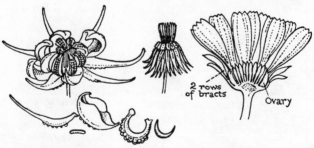

473.—FLOWER AND FRUIT OF MARIGOLD.

are often much more showy than the disc-florets, and they are sometimes not fertile. In the Linnæan system the relative fertility of the ray- and disc-florets provided a means of classification. Sometimes all the florets are tubular, with or without any increase of showiness at the margin. Tansy has a flower which seems to be robbed of its rays, while Corn-flower has its marginal tubular florets so much bigger than those of the disc that one does not at once recognize that they are of the same class. These marginal corollas in Corn-flower are without organs, and are merely for show, and their points are more than 5, and are moreover unequal in size. They thus approach Scabious.

Sometimes all the florets are ligulate, as in DANDELION, but in the most typical plants, as the DAISY, the disc-florets are small, tubular, densely crowded, and yellow in colour, and the ray-florets are ligulate, white, yellow, red or violet-blue.

The pappus of the calyx provides a means of dispersion, carrying the seeds upon the wind. Sometimes the pappus, or tuft of hairs or feathers, is close down upon the fruit, as in COLTSFOOT and GROUNDSEL, but sometimes it is borne upon a long shaft or tube, as in DANDELION and GOAT'S-BEARD. In COLTSFOOT the ray-florets only being fertile, the pappus is only produced at the ray, where it makes so white a frill to the flower as, at a little distance, to appear to be a white " daisy."

The pappus is, in some cases, represented by an inconspicuous membranous frill. It must be noted, as important in design, that the outer florets of the disc bloom before the inner ones. It is not uncommon, therefore, for there to be a ring of little stars—the open florets—at the edge of the disc, while the middle of the disc is occupied by little globules—smallest in the middle.

The most remarkable peculiarity of the Composites is the syngenesious character of their anthers. That s, while the filaments or stems of the 5 stamens are free and separate, the anthers are united edge to edge into a tube. This tube is yellow, brown, or purple, and plays an important part in the colour effect of some species. Up through the tube passes the style, which is often white, and which divides into 2 stigmas, usually curling gracefully to right and left, but subject to much variation. On account of their syngenesious anthers the Composites form, almost entirely, the Linnæan class, Syngenesia.

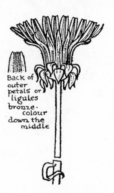

474.—HAWK-BIT.   475.—DANDELION.

The bracts, leaves, or scales of the involucres vary very considerably—from being in a single row, as in GOAT'S-BEARD, to being in several rows, as in THISTLES, TANSY, etc. Sometimes the bracts are leafy, green, and soft, sometimes hard and spiny, as in THISTLES, or are scaly with bristly edges, as in KNAPWEEDS. In DANDELION the lower bracts are usually reflexed. Sometimes the bracts continue for some distance down the stem, or perhaps the stem bears many scales, as in COLTSFOOT. The involucral bracts, whether green, purplish, or brownish, or light and gay in colour as in EVERLASTING FLOWERS, are generally shaded upwards into another colour. The ligular petals are not, as a rule, subject to shading from one colour to another, but some exotics, as Gaillardia, have a target-like appearance, and the common DAISY shades from white into pink. Often, however, as has been noted, the backs of the ligules differ in colour from the fronts. A similar colour-change is seen in CORN-FLOWER, where the tubular corollas of the disc are purple and the large ones of the ray are blue, white, or violet.

The leaves often form a rosette of leaves upon the ground, but sometimes are confined to the stems. They are alternate (except in a few cases such as Zinnia), without stipules, though they sometimes clasp the stem with ear-like projections, as in SOW-THISTLE. They are simple in general outline, usually spear-shaped or heart-shaped, often deeply lobed and toothed, but not serrated. It is not uncommon for some at least of the lobes to be directed backwards, as in Dandelion. In many cases the leaves are deeply cut into a pinnate form. They have no separate stalks, but the midrib serves the purpose, and often is bordered with a narrow fringe of blade. Indeed there is some similarity between these leaves and those of Crucifers.

Some further figures of plants of this Order are given in the Appendix.

477.—YELLOW GOAT'S-BEARD.
TRAGOPOGON PRATENSE. Z. 710.1.

476.—YELLOW GOAT'S-BEARD.
TRAGOPOGON PRATENSIS. F. 821.
Yellow. 18 in. Fl., 1½ in. June—July.

478.—PURPLE GOAT'S-BEARD. SALSIFY.
TRAGOPOGON PORRIFOLIUS. G. 595.1.
Purple. 3 ft. Fl., 1½ in. May—June.

479.—PURPLE GOAT'S-BEARD.
TRAGOPOGON PORRIFOLIUS. Z. 710.2.
Purple. 3 ft. Fl., 1½ in. May—June.

480.—OX-TONGUE.
HELMINTHIA ECHIOIDES. J. 798.2.
Yellow. 2–3 ft. Fl., ¾ in. June—July.

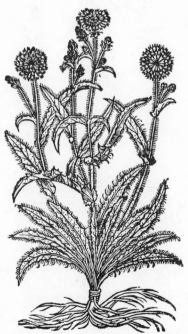

481.—YELLOW SUCCORY.
PICRIS HIERACIOIDES. J. 298.7.
Yellow. 2–3 ft. Fl., ¾ in. July—Aug.

482.—BLUE SOW-THISTLE.
SONCHUS ALPINUS. Z. 570.2.
Blue. 3–4 ft. July—Aug.

483.—COMMON SOW-THISTLE.

Sonchus oleraceus (asper). F. 675.

2–3 ft. Flowers pale yellow. Uncultivated ground, roadsides, hedges. May.

This plant differs from that in Fig. 484 in having its leaves deeply cleft.

484.—COMMON SOW-THISTLE.
Sonchus oleraceus (asper).   F. 674.

Fields, woods where the underwood has been cleared.   July—Nov.

485.—COMMON SOW-THISTLE.
SONCHUS OLERACEUS.
(See preceding figures.)

486.—MARSH SOW-THISTLE.
SONCHUS PALUSTRIS. J. 294.9.
Yellow. 7–8 ft. Fl., 1¼ in. July.

487.—DANDELION.
LEONTODON TARAXACUM. D. 564.1.
Bright yellow. 6 in. Fl., 1½ in. Ap.—Oct.

488.—ORANGE HAWKWEED.
HIERACIUM AURANTIACUM. J. 305.3.
Deep orange. 1 ft. Fl., 1 in. June—Aug.

489.—LETTUCE.
Lactuca sativa.   F. 299.
3–4 ft.   Flowers yellow.   June—July.
This may perhaps be *Lactuca crispa*, the Curled Lettuce.

490.—DANDELION.

LEONTODON TARAXACUM. F. 680.

Leaves runcinate (the points directed backwards), smooth, all springing from the root. Outer scales of the involucre reflexed (see Fig. 475). Height, 6 inches. Flower, 1½ in., bright yellow. Ap.—Oct. For another figure, see Fig. 487.

491.—AUTUMNAL HAWK-BIT.
LEONTODON AUTUMNALIS. F. 320.
Teeth of the leaves pointing, some forwards, some backwards. 6 in. to 2 ft. high. Flowers yellow.
Meadows, pastures. July—Aug. See Fig. 474.

492.—BROAD-LEAFED WALL
HAWKWEED.
Hieracium murorum. J. 304.1.
Deep yellow. 2 in.–2 ft. Fl., 1 in. Aug.

493.—MARSH SUCCORY-LEAFED
HAWKWEED.
Hieracium paludosum. J. 300.11.
Bright yellow. 3 ft. Fl., ¾ in. July—Aug.

494.—"YELLOW SUCCORIE." G. 222.2.

495.—"SUCCORIE HAWKWEED." G. 234.5.

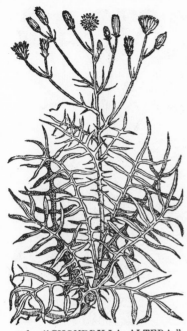

496.—"CHONDRILLA ALTERA."
M. 260.2.

497.—ENDIVE. CHICORY.
CICHORIUM INTYBUS. M. 258.2.
Bright blue. 1–3 ft. Fl., 1¼ in. July—Sep.

498.—"BLACK HAWKWEED."
G. 233.3.

499.—"BROAD-LEAFED MOUNTAIN
HAWKWEED." J. 299.10.

500.—"ROBINUS GUM SUCCORIE."
G. 224. 2.

501.—"YELLOW GUM SUCCORIE."
G. 225. 3.

502.—BURDOCK.
ARCTIUM LAPPA.   M. 502. 1.
Purple.   5 ft.   July—Aug.

503.—PURPLE SAW-WORT.
SERRATULA TINCTORIA.   G. 576. 1.
Purple.   2–3 ft.   Aug.—Sep.

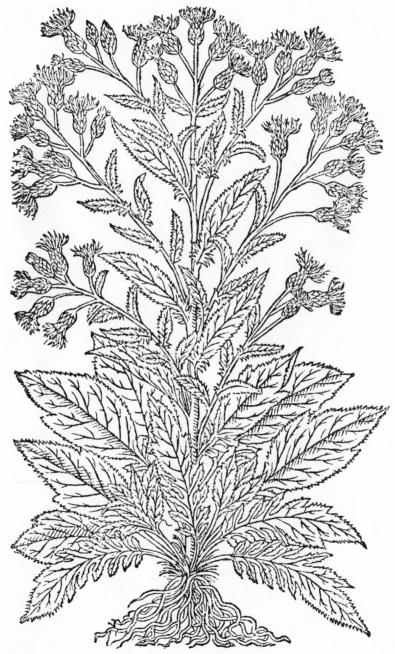

504.—COMMON SAW-WORT.

Serratula tinctoria. G. 577.3.

Florets purple. 2–3 ft. Aug.—Sep.

This drawing, which is enlarged, shows on its present scale the flowers about half-size.　Common in woods.

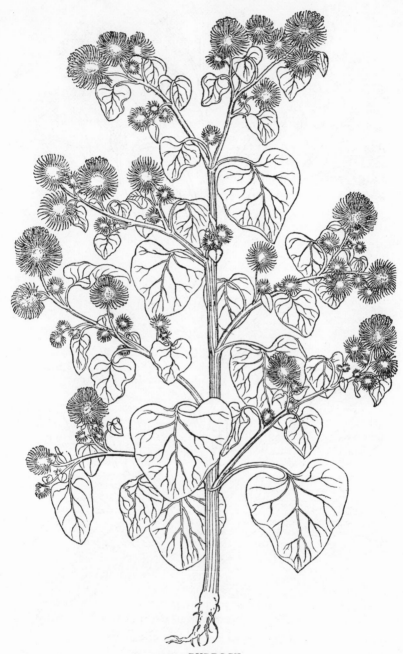

505.—BURDOCK.

ARCTIUM LAPPA. F. 71.

5 ft. Waste places. July—Aug.

This shows the upper two-thirds of the plant, the longer and larger lower leaves being omitted. Florets, purple, extending but little beyond the hooked bracts of the involucre.

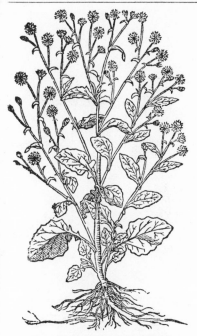

506.—COMMON NIPPLEWORT.
LAPSANA COMMUNIS. G. 231.8.
Pale yellow. 2–3 ft. Fl., ⅝ in. July—Aug.

507.—MUSK THISTLE.
CARDUUS NUTANS. D. 1472.2.
Deep purple. 3 ft. Fl., 2 in. July—Aug.

508.—MILK THISTLE.
CARDUUS MARIANUS. G. 989.
Purple. 5 ft. June—July.

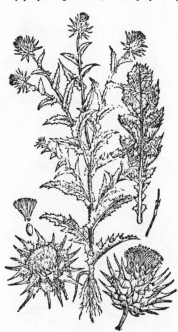

509.—OUR LADY'S, OR MILK
THISTLE.
CARDUUS MARIANUS. C. E. 445.
White veins upon its leaves.

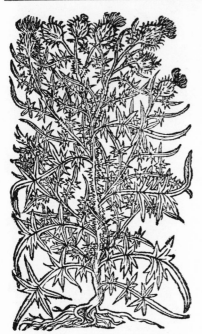

510.—SPEAR PLUME THISTLE.
CARDUUS LANCEOLATUS. J. 1174.6.
Purple. 3 ft. Fl., 2 in. long. July.

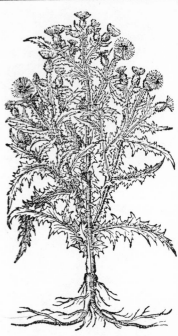

511.—WAY, OR CORN THISTLE.
CARDUUS ARVENSIS. L. i. 21.1.
Pale purple. 2 ft. July.

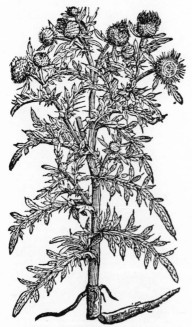

512.—WOOLLY-HEADED THISTLE.
CARDUUS ERIOPHORUS. L. ii. 9.2.
Purple or white. 5 ft. July.

513.—DWARF THISTLE.
CARDUUS ACAULIS. J. 1158.
Purple. 2 in. July.

514.—SCOTCH, or COTTON THISTLE.

ONOPORDUM ACANTHIUM. F. 57.

Covered with a white cotton, giving it a whitish-green colour. Stem with leafy, thorny borders. Leaves completely decurrent. 5 ft. Flowers large, purple; the calyx a depressed globe. Waste places. July—Aug. This is commonly called the Scotch Thistle, but the name is also given to Carduus lanceolatus, Fig. 510—the commonest of all thistles.

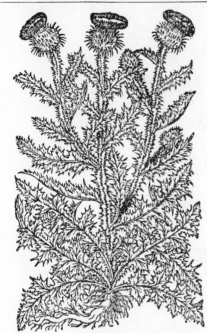

515.—SCOTCH THISTLE. G. 988.1.

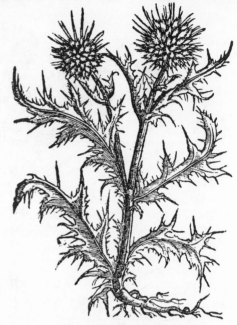

517.—"SPINA ALBA."
? ECHINOPS SPINOSA. L. ii. 9.1.
White. 4 ft. Egypt. July—Aug.

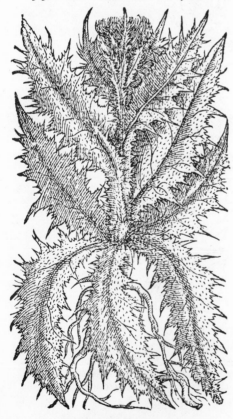

516.—SCOTCH THISTLE. D. 1446.

518.—"FLAT-HEADED GLOBE THISTLE." J. 1151.5.
ECHINOPS STRIGOSA has leaves like these—pale beneath,
but the flowers are globular.

519.—GLOBE THISTLE.
ECHINOPS SPHÆROCEPHALUS. F. 883.
5 ft.　Flowers light blue.　Leaves downy above, woolly beneath.　Austria.　July—Aug.

520.—"GLOBE THISTLE."
ECHINOPS. G. 990.
Probably the same as Fig. 519.

521.—WAY, CORN, OR CREEPING THISTLE.
CARDUUS ARVENSIS. J. 1173.4.
Pale purple. 2 ft. July.

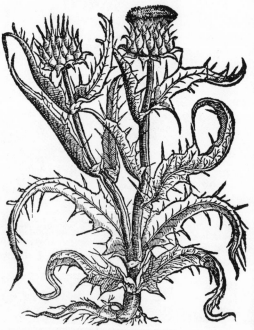

522.—WELTED THISTLE.
CARDUUS ACANTHOIDES. D. 1473.
Purple. 2 ft. July.

523.—ARTICHOKE.
CYNARA SCOLYMUS. J. 1153.3.
Purple. Head 4 in. S. Europe. Aug.—Sep.

524.—ARTICHOKE.
CYNARA SCOLYMUS. F. 792.

The plant is here shown before the stem is developed. For the complete plant see the preceding and following ~~fig~~ures. The plants are covered with litter during winter, and are uncovered and the ground dressed in March or April ~~in~~ June the heads appear. The lower parts of the leaves of the involucre, the fleshy receptacles of the flowers—freed ~~fro~~m the bristles and " choke "—are boiled as a vegetable.

525.—CARDOON.
CYNARA CARDUNCULUS. D. 1438.2.

The Cardoon is an Artichoke. After flowering in July it is cut down and allowed to grow again. By September it will have leaves 2 ft. long. These are tied round with straw, and earthed. The stalks of the inner leaves are thus blanched and rendered crisp and tender, like celery, and are used in the same way as it (see Loudon). Fig. 524 may represent a plant ready for earthing. These Artichokes must not be confused with the Jerusalem Artichoke, which is a North-American Sunflower, Helianthus tuberosus, having sweet edible tuberous roots. "Jerusalem" is a corruption of the Italian *girasole* = turning to the sun.

CARLINE THISTLE.—A kind of Everlasting Flower, the bracts of the involucre forming rays around the florets, and appearing to be petals. Height up to 15 in. Florets red or purple, straw-coloured below. Involucre with "scales purplish-edged and terminated with branching yellow thorns: the innermost, strap-shaped, pointed at the end; dry, straw-coloured within; without, reddish brown towards the base." The involucre being permanent, the plant remains unaltered in general appearance for some time. "It continues, after it is dead, unchanged even for the whole of the second year, a mournful spectacle."—Linnæus.

526.—CARLINE THISTLE.
CARLINA VULGARIS. D. 1439.2.
Red. 3 in.–1 ft. Fl., 1¼ in. June—July.

527.—NODDING BUR MARIGOLD.
BIDENS CERNUA. G. 574.
Yellow. 1–2 ft. Fl., 1¼ in. June—Sep.
The "rays" are bracts. Florets all tubular.

528.—"SMALL BURRE THISTLE."
D. 583.3.

529.—"BROAD-LEAFED GOLDILOCKS."
J. 646.2.
? A species of Chrysocoma.

530.—SEA COTTON-WEED.
DIOTIS MARITIMA. G. 516.3.
Yellow. 1 ft. Fl., ⅝ in. Aug.—Sep.

532.—TANSY.
TANACETUM VULGARE.
Yellow. 1–3 ft. Fl., ¾ in. July—Aug.

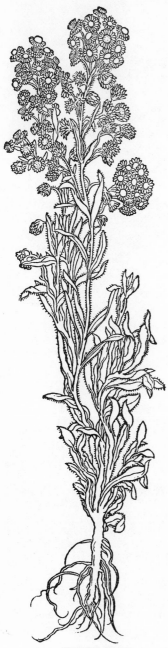

533.—MUGWORT.

The full plant is shown on the opposite page. The flowers are of oval form as figured in this detail. The little stipule-like appendages to the leaves should be noted. Such are very often present in the composites.

531.—GOLDILOCKS.
ASTER LINOSYRIS, OR LINOSYRIS VULGARIS. F. 99.

1 ft. Flowers yellow, without rays in England ; a variety with rays (says Bentham) in Germany. Rocky places. Aug.—Sep.

Just as some Composites are spiny and hard in the leaf, so others are soft and downy. Some oɪ the Thistles are cottony, notably the SCOTCH or COTTON THISTLE. It is a common thing, therefore, for some part of the plant to be white, in consequence of the down or cotton upon it. Often the under-sides of the leaves are white. This gives an excellent colour-change. Of WORMWOOD the leaves are dark green on the upper, white on the under surface, and are silky and very soft. The leaves of Mugwort are similar, and the buds and flowers of it are, in the same way, silvery white. Many of the plants which follow have this cottony character. The backs of the leaves of COLTSFOOT are very white—the veins reddish—and make a conspicuous display when turned about by the wind.

534.—COMMON WORMWOOD.
ARTEMISIA ABSINTHIUM. M. 329.1.
Yellow. 1 ft. Aug.

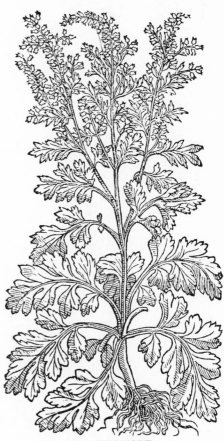

535.—MUGWORT.
ARTEMISIA VULGARIS. M. 399.
Reddish or yellow. 4 ft. Aug.

536.—HERB IMPIOUS. CUDWEED.
Gnaphalium germanicum. G. 517.9.
Pale yellow. 9 in. Heads globose. June—July.

537.—CAT'S FOOT.
Antennaria dioicum. G. 516.6.
White or rose. 6 in. June—July.

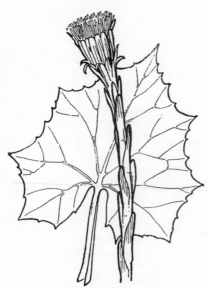

538.—COLTSFOOT.
Tussilago Farfara.
Yellow. Scales pink. 4–12 in. Mar.—Ap.;
before the leaves. The flower here shown closed.
Stemless.

539.—BUTTERBUR. BOY'S HAT.
Tussilago Petasites.
Pale lilac. 6 in. Ap.—May.
Flowers appear before the leaves.
Leaves very large, downy beneath.
(See Frontispiece and next page.)

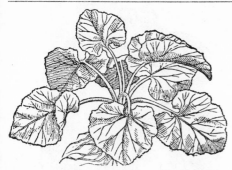

BUTTERBUR. M.

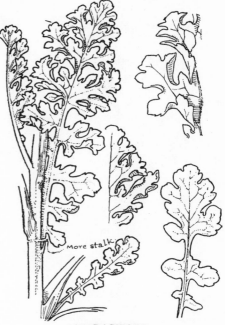

540.—RAGWORT.
Details of leaves.

RAGWORT.—Few plants excel the different species of Ragwort in their gayness. Fig. 542, good as it is, does not quite express the growth of the plant. It is deficient in the strong swift lines of the branches, which should rise up stiffly and expand above into a broad mass of flowers, not into a few only.

Says Gerarde—"Saint James his woort or Ragwoort is very well known everywhere, and bringeth foorth at the first broade leaves, gashed round about . . . not whitish or soft, of a deepe greene colour, with a stalk which rises up about a cubite high, chamfered, blackish, and somwhat red withal."

Of Fig. 544 he said—"The flowers vanish into downe, and fly away with the winde." He says it has ash-coloured stems, and was therefore called Cineraria.

541.—BROAD-LEAFED RAGWORT.
SENECIO SARACENICUS. G. 350.
Yellow. 3–5 ft. Fl., 1 in. July—Aug.

542.—RAGWORT.
SENECIO JACOBÆA. G. 218.1.
Deep yellow. 2–3 ft. Fl., 1 in. July—Aug.

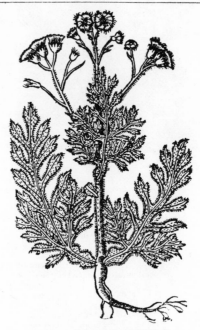

543.—GROUNDSEL.
SENECIO VULGARIS.
Yellow (without rays).　4-12 in.　All the year.

544.—"SEA RAGWEED."
J. 280.4.

545.—BIRD'S TONGUE GROUNDSEL.
SENECIO PALUDOSUS.　J. 483.6.
Yellow.　5 ft.　Fl., 1¼ in.　June—July.

546.—SEA STARWORT.
ASTER TRIPOLIUM.　D. 1389.
Rays, blue or purple ; disc, yellow.　6 in.–3 ft.
Muddy sea-shores.　Aug.—Sep.

547.—" ASTER ATTICUS ALTER."   M. (1565) 1176.

Asters are so named from the star-like arrangement of the leaves of the involucre. In colour they are usually purple, blue, red, violet, or white. The double varieties have the outer ligules richer in colour than the inner. All possible tints of the above-mentioned colours are to be found ; but yellow seems to be very unusual. The plant here figured was said to be yellow. Michaelmas Daisy and Christmas Daisy are names given to species which flower at those times. Usually the disc is yellow or green, and the ray violet.

548.—"CREEPING STARWORT."
D. 861.2.

549.—"SALLOW-LEAFED STARWORT."
J. 488 9.

550.—ITALIAN STARWORT.?
ASTER AMELLUS. M. 510.
Purple. 2 ft. Aug.—Sep.

551.—GOLDEN ROD.
SOLIDAGO VIRGAUREA. G. 348.2.
Yellow. 1–3 ft. Fl., ½ in. July—Sep.

552.—ELECAMPANE.
INULA HELENIUM.   M. (1565) 71.

Stem 5 or 6 ft. high, branched towards the top, scored, cottony.   Leaves, the lower on leaf-stalks, serrated or toothed, deep green, slightly hairy above, whitish green and thickly cottony underneath.   Flowers large, 2 in. or more across, yellow, solitary, terminating the stem and branches.   The rays are fully an inch long, terminating in 3 sharp points, of which the middle one is the shortest.   This detail is not expressed here, but is given in Gesner's cut—the next figure.   Moist meadows and pastures.   July—Aug.

553.—ELECAMPANE.
INULA HELENIUM. C.E. 35.
(See the previous figure.)

554.—PLOUGHMAN'S SPIKENARD.
INULA CONYZA. M. 405.1.
Dingy yellow. 1 ft. Aug.—Sep.

555.—GREAT LEOPARD'S BANE.
DORONICUM PARDALIANCHES. G. 620.1.
Yellow. 2-3 ft. Fl., 2 in. June—Sep.

556.—FLEA-BANE.
ERIGERON GRAVEOLENS? M. 405.2.
Yellow. 18 in. S. Europe. Aug.

557.—FLEA-BANE.

INULA, OR PULICARIA, DYSENTERICA. F. 436.

Apparently drawn from a gathered specimen which has drooped. 1–2 ft. Stem and leaves woolly. The plants grow together, and have a sturdier appearance than this figure presents. Flowers deep yellow, the rays numerous and narrow. Ditches and moist places. Aug.—Sep.

558.—DAISY.

BELLIS PERENNIS. F. 146.

Rays white, notched at the end, tipped with crimson, and pink underneath. Involucre of from 10 to 20 narrow bracts in a double row. Florets of the disc yellow, sometimes converted into ligules and producing the double variety, as is represented in the right side of the above figure. When wild the plant makes a close tuft, with all the leaves radical ; but in cultivation stems are formed, as here shown. Mar.—Sep.

559.—"MIDDLE WILD DAISY."
J. 636.5.

560.—"WHITE OXE EYE."
G. 607.3.

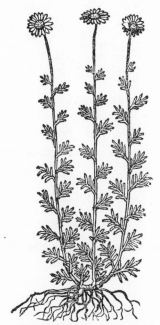

561.—"YELLOW OXE EYE."
"BUPHTHALMUM." M. 418.
Matthiolus says the colour is yellow.

562.—CORN MARIGOLD.
CHRYSANTHEMUM SEGETUM. J. 743.1.
Yellow. 1–2 ft. Fl., 1¼in. June—Aug.

563.—OX-EYE DAISY.
CHRYSANTHEMUM LEUCANTHEMUM. G. 509.
White, disc yellow. 2 ft. Fl., 1¾ in. June—July.

565.—CHRYSANTHEMUM.
CHRYSANTHEMUM CORONARIUM? M. 465.
Yellow. 4 ft. Sicily. July—Sep.

The many handsome varieties in the garden are derived from C. indicum and C. sinense, the Indian and Chinese kinds. They are white, yellow, pink, red, violet.

This plant is very like the winter-flowering Marguerite—Chrysanthemum frutescens, a species from Teneriffe, with a yellow disc and white rays. The name Marguerite is a common French word for Daisy, particularly of the common field daisy and of the ox-eye daisy.

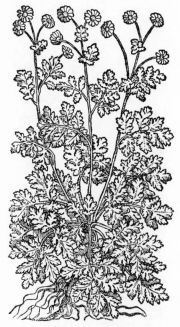

564.—FEVERFEW.
CHRYSANTHEMUM PARTHENIUM. M. 417.2.
White, disc yellow. 1–2 ft. Fl., ⅝ in. July.

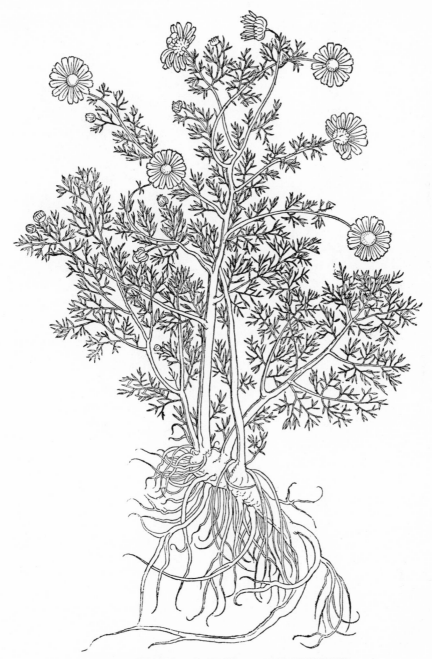

566.—SCENTLESS MAYWEED. CORN FEVERFEW.
MATRICARIA INODORA. F. 144.

Distinguished from Anthemis Cotula and Anthemis arvensis by want of chaff on the receptacle, and from Matricaria Chamomilla by its flattened involucre. 1 ft. Fl., 1¼ in. Cornfields, roadsides. July—Sep.

567.—CHAMOMILE.
MATRICARIA CHAMOMILLA. F. 25.

Rays white, disc yellow—the receptacle conical.   1 ft.   Flowers ¾ in., solitary, terminating the branches.
May—Aug.

568.—WILD CHAMOMILE.
Matricaria Chamomilla. M. 417.1.
White, disc yellow. 1 ft. Fl., ¾ in. May—Aug.

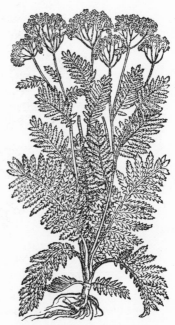

569.—YARROW. MILFOIL.
Achillea Millefolium. M. 498.1.
White. 1–2 ft. Fl., ¼ in. June—Aug.

Milfoil.—Fig. 569 very well represents the foliage, but not the flowers. The heads are flatter, and the flowers themselves more visible. They are daisy-like, but with only small middles, and about five rays, which are broad, notched at the end, and creased with two furrows. There are two or three other kinds wild in England : among them is the Golden Milfoil with yellow flowers, and another with pale yellow flowers.

Blessed Thistle is so named from its supposed extraordinary medical properties.

Hemp Agrimony or Hemp-weed, Eupatorium cannabinum, is a common and handsome plant which must be mentioned. It rises to about 3 ft., upright, with opposite pinnate leaves, sharply serrated. The leaves have 3 or 5 long segments joining together so close to the stem as not readily to be followed. On examination they remind one of Dahlia leaves. The florets are about 6 together in a flower, which is of the long, narrow kind, as in Mugwort (see Fig. 533). The flowers are in dense, terminal, corymbose heads, suggesting Valerian. They are pink, and rendered more conspicuous by the long projecting styles and stamens. July—Aug.

570.—BLESSED THISTLE.
Centaurea benedicta. Z. 775.
Yellow. 2 ft. June—Oct.

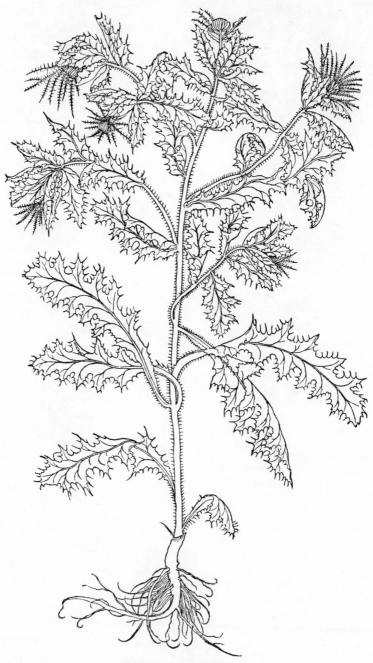

571.—HOLY or BLESSED THISTLE.
CENTAUREA BENEDICTA. F. 122.
Yellow. 2 ft. Spain. June—Oct.

572.—BLACK KNAPWEED.
CENTAUREA NIGRA. G. 588.1.
Purple. 2–3 ft. Fl., 1¼ in. high. June—Sep.

573.—"GREAT SILVER KNAPWEED."
D. 1109.2.

574.—VIOLET-COLOURED BOTTLE.
CENTAUREA CYANUS. G. 593.5.
Blue. 2–3 ft. Fl., 1½ in. July—Sep.

575.—BLUE-BOTTLE. CORNFLOWER.
CENTAUREA CYANUS. L. i. 546.2.
Outer florets blue, inner florets purple.

576.—"KNOBBIE KNAPWEED."
G. 589.6.

577.—COMMON STAR-THISTLE.
Centaurea Calcitrapa. J. 1166.1.
Pink. 1–2 ft. Fl., 1 in. July—Sep.

578.—ST. BARNABY'S THISTLE.
Centaurea solstitialis. J. 1166.2.
Yellow. 2 ft. Fl., ¾ in. June—Oct.

579.—LESSER BURDOCK.
Xanthium strumarium. C. E. 926.
Greenish. 1–2 ft. Aug.—Sep.

580.—MARIGOLD.
CALENDULA OFFICINALIS. G. 600.1.

581.—MARIGOLD.
CALENDULA OFFICINALIS. J. 739.3.
Orange. 3 ft. Fl., 2½ in. June—Sep.

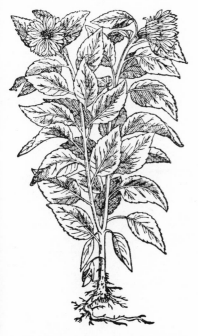

582.—PERENNIAL SUNFLOWER.
HELIANTHUS MULTIFLORUS. G. 613.4.
Yellow. 6 ft. Fl., 4 in. Aug.—Oct.

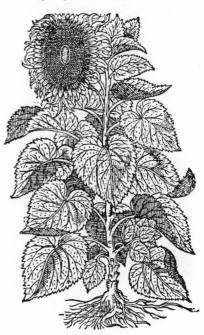

583.—SUNFLOWER.
HELIANTHUS ANNUUS. G. 612.1.
Yellow. 6 ft. Fl., 10 in. June—Oct.
(See Appendix, Fig. 29.)

584.—AFRICAN MARIGOLD.

TAGETES ERECTA.　D. 840.

Yellow.　3 ft.　June—Sep.

The petals sometimes longitudinally variegated—striped, with rich red-brown. This figure is an example of the admirable re-drawing which was occasionally done. For the figure, though copied from Fuchsius, and remarkably like his small figure, is not traced from it. Looking very closely into the details, one sees that some of the strong feeling for boldly-turned form which distinguished the German work is lost in this French rendering. It was executed fifty years later, however ; and some decline had taken place throughout the whole of graphic art.

FRENCH MARIGOLD—Tagetes Patula.—Called also the Spreading Marigold. The rays are about 8. They are broad-notched, somewhat fish-tail shaped, and often variegated with longitudinal stripes. Sometimes the ray presents three stripes, the middle one yellow, the outer ones red. The back of the flower is free from bracts, the involucre being rather a one-leafed, five-toothed calyx, tubular, with a narrow mouth. The stem thickens up to the calyx, and is deeply ridged in the same way as the calyx.

ZINNIA.—A familiar garden flower, similar in some respects to the French and African Marigolds. Its stem also thickens up to the flower ; but the calyx has imbricated scales, and thus is properly an involucre. The rays are broad, slightly notched, and boldly modelled. They bend back and expose the disc florets. The rays are generally red or scarlet, but in some species yellow and purple. The leaves in this genus are opposite, and incline one at first sight to suspect the plants of belonging to the Teazel Order.

SUNFLOWER—Helianthus annuus.—This splendid flower was, on its first introduction into Europe, called the Marigold of Peru, and the Great Golden Sun. An engraving of it by Crispin de Passe is given in the Appendix, Fig. 29. The leaves of this species, as of the small perennial Sunflower, Helianthus multiflorus, and of the Jerusalem Artichoke, Helianthus tuberosus, are 3-nerved, a peculiarity important, and useful, in design.

DAISY TREE—Olearia.—The composites are rarely more than herbs, but some Asters are shrubby, and there are some shrubs of the genus Olearia which are common in gardens, or rather, in shrubberies and parks. The most frequent has small flowers not unlike those of Milfoil, and are not immediately seen to be daisies—Olearia Hastii, which is only 3 ft. high ; but Olearia Gunniana has more daisy-like flowers, and reaches to 8 ft.

CAMPANULACEÆ—BELL-FLOWERS.

Herbs and undershrubs with a white milky juice. Leaves almost always alternate, simple, or deeply divided, without stipules. Flowers single, in racemes, spikes, or panicles, or in heads; usually blue or white, sometimes rose-violet, rarely yellow. Calyx superior, usually 5-lobed (3–8), persistent. Corolla, monopetalous

inserted into the lip of the calyx (its outline not continuous with that of the calyx), usually 5-lobed (3–8) withering on the fruit. Lindley: "Stamens usually 5, anthers usually distinct, not, as in Lobeliads, joined together. The anthers are long, upon short, broad filaments. Style thick, hairy, the stigma simple or lobed with, perhaps, 5 lobes."

(For a figure of a species of Phyteuma, see Appendix, Fig. 15.)

The Lobeliads, of which many species are common in gardens, are sometimes placed in this Order. They have irregular corollas, and are tropical or sub-tropical plants.

585.—CLUSTERED BELL-FLOWER.
CAMPANULA GLOMERATA. L. i. 326.2.
Purple-blue. 6 in.–1 ft. Fl., ⅝ in. July—Aug.

586.—"SMALL COVENTRY BELLS."
G. 364.4.

587.—PEACH-LEAFED BELL-FLOWER.
CAMPANULA PERSICIFOLIA. L. i. 327.1.
Blue. 2ft. Fl., 1 in. July.

588.—CANTERBURY BELLS.
CAMPANULA MEDIUM. L. i. 324.
Blue or white. 4 ft. June—Sep.

589.—HAREBELL.
CAMPANULA ROTUNDIFOLIA.
Blue. 8 in.–2 ft. Fl., $\frac{3}{4}$ in. long. July—Sep.

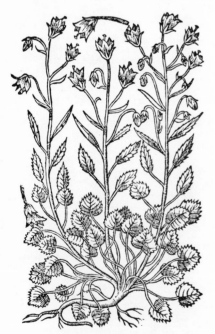

590.—HAREBELL.
Z. 461.1.

591.—STEEPLE BELLS.
CAMPANULA LATIFOLIA. J. 448.3.
Blue, pink, or white. 4 ft. Fl., 1 in. long. Aug.

VACCINIACEÆ—BILBERRY AND CRANBERRY.

Shrubs with alternate undivided leaves, without stipules, and succulent fruit. Calyx superior, corolla 4- to 6-parted, stamens twice as many. Style and stigma simple.

The CRANBERRY is Vaccinium Oxycoccus, a slender trailing shrub, 2–4 in. only. Peduncles 1-flowered, terminal. Flowers pink, berries red. June.

BOG WHORTLEBERRY, V. uliginosum, is 1 ft. high, with pink flowers and black berries.

592.—BILBERRY. WHORTLEBERRY.
VACCINIUM MYRTILLUS. Z. 84.
Pink. Berries black. 2 ft. May.

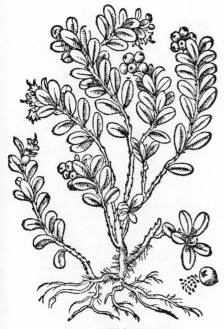

593.—COW-BERRY.
VACCINIUM VITIS IDÆA. Z. 85.1.
Pink. Fruit, red. 9 in. May.

ERICACEÆ—HEATH AND RHODODENDRON.

Shrubs. The Order is represented in Britain only by the Heathers, but it includes such handsome shrubs as Rhododendron, Azalea, Andromeda, Kalmia, etc. (See Appendix, Figs. 53 and 54.)

There is a 4- or 5-cleft, persistent, calyx, and a more or less bell-like, or egg-like, corolla, parted at the margin into 4 or 5 points. Stamens 8. Style 1. Ovary superior.

The leaves are evergreen, rigid, entire, whorled or opposite, and without stipules.

[This Order begins those of the *Corollifloræ* which have the ovary above the calyx.]

594.—COMMON HEATH.
ERICA CINEREA.
Purple-red. July—Oct.

595.—LING. HEATHER.

Calluna vulgaris. F. 255.

The calyx is 4-parted coloured, and covers the corolla, which is smaller, and also 4-parted. At the base of the calyx are 4 bracts which look like a calyx at first sight. Leaves imbricated, in 4 rows. Height, 6 in. to 4 ft. Flowers pink. June—Oct.

596.—FINE-LEAFED HEATH.
Erica cinerea. J. 1382.7.
Purple-red.   6 in.–2 ft.   Leaves 3 together.   July—Oct.

597.—STRAWBERRY-TREE.
Arbutus Unedo. Z. 96.2.
White or pink.   10 ft.   Ireland.   Sep.—Oct.

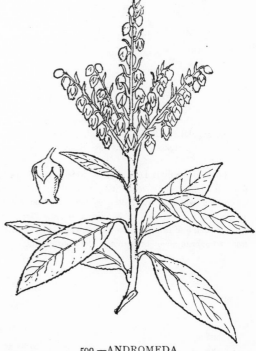

598.—STRAWBERRY-TREE.
Arbutus Unedo. L. ii. 141.1.
Berry red.   ½ in.

599.—ANDROMEDA.
Various species are common in shrubberies.   Flowers, white
or pink, ½ in. long.   Leaves 5 in., sometimes rusty.   3–20 ft.

PYROLACEÆ.

Herbs or undershrubs, represented in Britain only by the Winter-Greens, which are plants of from 6 to 10 in. in height, with simple evergreen leaves. Sepals 5, inferior. Petals 5, but slightly united. Style 1. Stamens 10.

600.—WINTER-GREEN.
PYROLA ROTUNDIFOLIA. F. 467.
White. 10 in. Fl., ⅝ in. July—Aug.

EBENACEÆ—EBONY.

Trees or shrubs with entire, leathery, alternate leaves, without stipules. Calyx 3–7-parted, inferior, persistent. Corolla 3–7-parted. Stamens as many, or in some multiple.

The chief genus is Diospyros, of which several species produce Ebony of different kinds, and others yield other ornamental woods. Some bear edible fruit, notably D. Virginiana, the Persimmon, or Virginian Date-Plum, and D. Lotus, the European Date-Plum, which is one of the fruits thought to have been eaten by the Lotus-eaters.

601.—EUROPEAN DATE-PLUM.
DIOSPYRUS LOTUS. D. 349.2.

602.—EUROPEAN DATE-PLUM.

DIOSPYROS LOTUS. M. (1565) 257.

Fruit the size of a Cherry, yellow when ripe.

Diospyrus = The Fruit of Jove. This is one of the plants supposed to cause oblivion, the Lotus. 20 ft.

AQUIFOLIACEÆ—HOLLY.

Evergreen trees or shrubs whose branches are often angular. Leaves alternate, or opposite, simple, leathery, without stipules. Flowers small, white or greenish, axillary, solitary or clustered. Sepals 4–6. Corolla 4–6-parted, beneath the ovary. Stamens alternate with the segments of the corolla. Style simple. Fruit fleshy, with 2–6 or more stones.

The Holly, as a tree, reaches 30 to 40 ft. in height, but is more common as a hedge bush, or low tree. The flowers cluster handsomely upon the branches in June.

WINTERBERRY, of which there are deciduous and evergreen kinds, Prinos verticillatus and Prinos glaber are shrubs from 6 to 12 ft. high. They have alternate spear-shaped leaves of fair size. The flowers appear in July, in the axils of the leaves, 2 or 3 together on a foot-stalk. They make no show, but are succeeded by red and purple berries which remain on the trees all the winter, and have a good effect. Both are of North America.

603.—HOLLY.
ILEX AQUIFOLIUM. L. ii. 153.2.
White. Berries scarlet. 6–20 ft. Br. May.

STYRACACEÆ—STORAX.

Shrubs or trees, with alternate leaves without stipules. Calyx with 5 or 4 divisions. Corolla with 5–10 petals. Stamens about 10. Style simple. Fruit more or less fleshy.

Storax and Benzoin, fragrant gum resins, are the produce of two species of Styrax.

Halesia is the SNOWDROP TREE.

604.—STORACE OR STORAX.
STYRAX OFFICINALE. Z. 169.
White. 12–20 ft. Syria. July.

APOCYNACEÆ—DOGBANES.

This Order consists largely of exotic shrubs and trees, generally poisonous, or suspected of being so, though some have edible fruits or nutritive juices. Tabernæmontana dichotoma is said to be the forbidden fruit of Paradise. The fruit has much beauty, and the flower is fragrant. The fruit still bears the marks of Eve's teeth ! But the delicious fruit is now poisonous !

The flowers in this Order have a 5-parted, persistent, calyx. The corolla is 5-lobed, often with scales in the throat. Stamens 5, upon the corolla. Styles 2 or 1. Stigma 1, "assuming much the appearance of an hourglass."—Lindley.

The only British species are the Lesser and Greater Periwinkle. The latter has larger, ovate, leaves, and a flower an inch and a half across. It flowers in May and June. Both kinds are evergreen.

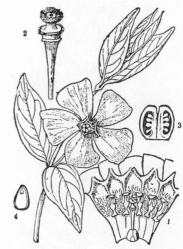

605.—GREATER PERIWINKLE.
VINCA MAJOR.　L. V. K. 599.
Purplish-blue. Trailing. Fl., 1¼ in. May—June.

606.—LESSER PERIWINKLE.
VINCA MINOR.　F. 360.
Blue. 6 in. Fl., 1 in. Apr.—July.

607.—DOUBLE PERIWINKLE.
VINCA MINOR FLORE PLENO.　C. E. 695.

608.—ROSE BAY OLEANDER.
NERIUM OLEANDER. G. 1220.1.
Red. 8 ft. June—Oct.

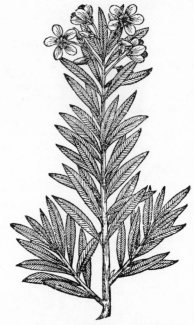

609.—WHITE-FLOWERED OLEANDER.
NERIUM OLEANDER. G. 1220.2.
Grows beside rivers and torrents.

GENTIANACEÆ—GENTIANS.

These plants differ from Dogbanes chiefly in being herbaceous, and in the leaves being more strongly ribbed. The corolla is mono-petalous, regular, with the same number of limbs as the calyx, 5, 4, 6, 8, or 10. Stamens the same in number. Style, 1 ; stigmas, 2.

The leaves are opposite entire, without stipules, sessile, sometimes confluent, with 3 or 5 ribs. Ovary superior.

FRINGED WATER LILY, Villarsia nymph-æoides, has a yellow flower, about 1 in. across, and heart-shaped leaves. It grows in pools and slow rivers. August. (See Appendix, Fig. 54.)

610.—MARSH GENTIAN.
GENTIANA PNEUMONANTHE. G. 355.
Blue, streaked green. 8 in. Fl., 1¼ in. long. Aug.

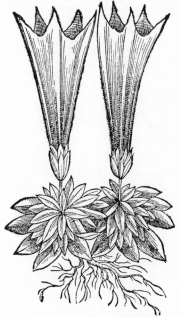

612.—DWARF GENTIAN.
GENTIANA ACAULIS. D. 828.2.
Deep blue. 6 in. Fl., 1½ in. long.
May—July.

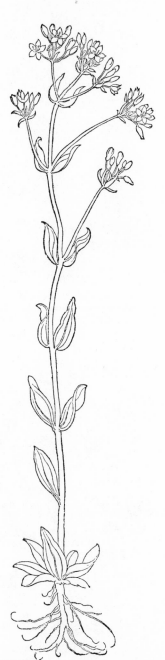

611.—LESSER OR COMMON CENTAURY.
ERYTHRÆA CENTAURIUM. F. 387.
Pink. 1 ft. Fl., ⅜ in. June—Sep.

613.—GREAT GENTIAN.
GENTIANA LUTEA. D. 1258.
Yellow. 4 ft. June—July.

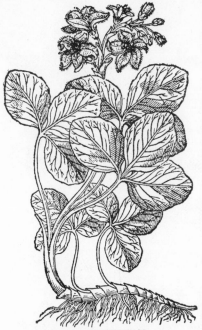

614.—CROSSWORT GENTIAN.
GENTIANA CRUCIATA.　F. 420.
Dark blue.　1 ft.　June—July.

615.—BUCKBEAN.
MENYANTHES TRILOBA.　G. 1024.1.
White, tipped pink.　8 in.　Fl., ¾ in.　June.

616.—YELLOW-WORT.
CHLORA PERFOLIATA.　C. E. 427.
Bright yellow.　1 ft.　Fl., ¾ in.　July—Sep.

617.—GREAT BINDWEED [CONVOLVULACEÆ].
CONVOLVULUS SEPIUM.　J. 861.1.
White or pinkish.　Fl., 3 in.　July—Aug.

CONVOLVULACEÆ—BINDWEED.

Herbs, or shrubs, usually twining. Leaves alternate; undivided or lobed, seldom pinnatifid; without stipules. Calyx persistent, with 5 deep divisions folded upon each other. Corolla 5-lobed, plaited, regular, deciduous. Stamens 5; style 1, or divided into several.

The CUSCUTACEÆ, or Dodders, are considered to be a sub-division of the Convolvulaceæ. They are parasites, have no leaves, and have scales in the throat of the corolla which Convolvulaceæ proper have not.

618.—SMALL BINDWEED.
CONVOLVULUS ARVENSIS.
Pale pink. 9 in. Fl., 1¼ in. June—July.

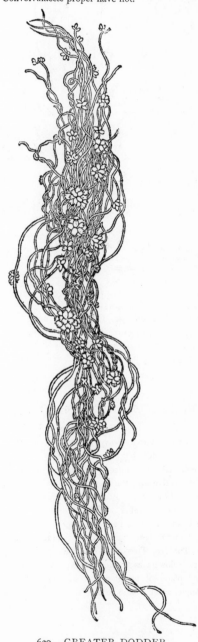

619.—GREAT BINDWEED.
CONVOLVULUS SEPIUM.
White or pinkish, with 5 broad yellowish veins. Calyx within 2 heart-shaped bracts. Fl., 3 in. July—Aug.

620.—GREATER DODDER.
CUSCUTA EUROPÆA. F. 348.
Reddish. 2 ft. Parasitic on Thistle and other plants. Aug.

321

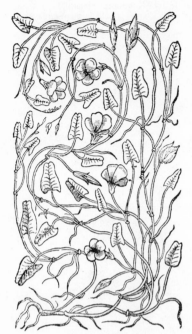

621.—SEA-COLEWORT [Convolvulaceæ].
Convolvulus Soldanella. D. App. 21.
Rose. Not climbing. Fl., 2 in. July.

The flower very like that of the Great Bindweed.
The leaves should be more kidney-shaped.

SOLANACEÆ—NIGHTSHADE.

An important Order of narcotic or poisonous
plants, having alternate leaves of angular outline,
generally undivided, but sometimes lobed.

The flowers are trumpet-like, suggesting at once
those of Gentian and Bindweed. The calyx and
corolla are both 5- or rarely 4-parted. The stamens
are the same in number, upon the corolla. One
stamen is sometimes sterile.

The Order includes such important plants as the
Potato, Tobacco, Tomato, Winter-Cherry, Capsicum ; such flowers as Petunia (see Appendix, Fig.
31), and plants of peculiar interest, such as the
Mad-Apple, and the Mandrake.

622.—COMMON HENBANE.
Hyoscyamus niger. B. 224.
Dingy yellow, veined purple. 1 ft. Fl., 1 in. July.

623.—"ENGLISH TOBACCO."
NICOTIANA RUSTICA. M. (1565) 1066.

So identified by Stokes. Loudon gives the plant as having green flowers, a height of 3 ft., and flowering in July to September. Hardier than N. Tabacum. Cultivated in cool latitudes.

624.—COMMON WINTER-CHERRY.
PHYSALIS ALKEKENGI. F. 687.

CAPE GOOSEBERRY, Physalis peruviana, is much the same as this plant, but its bladdery calyx is dull yellow, whereas the bladder of this is red. The present plant is native of the South of Europe, and its berries are eaten as a common fruit. Those of Cape Gooseberry, however, are preferable. So great a prejudice have botanists against the fruits of this Order that some have thought the use of Tomatoes, Winter-Cherries, etc., harmful, even when accepted and established as articles of food. These plants are 12 to 18 in. high, bear white flowers, which bloom—Alkekengi from July to September, peruviana from April to October.

625.—THORN-APPLE.
DATURA STRAMONIUM. C. E. 176.
White. 6 in.–2 ft. Fl., 1½ in. July—Sep.

626.—DOWNY THORN-APPLE.
DATURA METEL. C. E. 175.
White. 2 ft. Fl., 6 in. long. June—Sep.

627.—DEADLY NIGHTSHADE.
ATROPA BELLADONNA. Z. 1082.1.
Dull purple. Berries black. 3 ft. Fl., ⅝ in.
Br. June.

628.—WOODY NIGHTSHADE. BITTER-SWEET.
SOLANUM DULCAMARA. M. 553.
Bright purple. Stamens yellow. Berries red.
Fl., ⅝ in. June.

629.—CAPSICUM ANNUUM.　Z. 222.2.
White.　1 ft.　June—July.

630.—CAPSICUM ANNUUM.　D. 632.3.
Fruit at first green, then red.

631.—TOBACCO.
Nicotiana Tabacum.　J. 357.2.
White or pink.　4 ft.　Fl., 4 in. long.
July—Aug.

632.—BLACK WINTER-CHERRY.
D. 598.

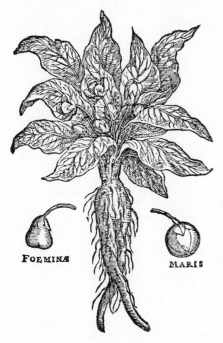

FOEMINA MARIS

633.—MANDRAKE.
ATROPA MANDRAGORA. L. i. 267.2.
White. 3 ft. Fruit yellow. Mar.—Apr.

634.—MAD-APPLE.
SOLANUM INSANUM. L. i. 268.1.
Blue. 2 ft. Fruit blue or white. July.

MANDRAKE.—Perhaps no plant has a greater hold upon the imagination than this. The name, built up of two common words, adds, there can be no doubt, to the belief or superstition which is connected with it. The ancient name was Mandragoras or Mandragora, which readily got corrupted into Mandrake or Man-dragon.

William Turner, in his *Herbal*, took pains to explode the popular superstitions regarding this plant. The Mandrake is poisonous and narcotic : it is said to have been used as an anæsthetic by the ancients. Gerarde says : "There have been many ridiculous tales brought up of this plant, whether of olde wives or some runnagate surgeons or phisickmongers, I know not, (a title bad enough for them:) but sure some one or moe that sought to make themselves famous in skillfull above others were the first brochers of that error I spake of. They adde further, that it is never or very seldom to be founde growing naturally but under a gallows, where the matter that hath fallen from the dead bodie, hath given it the shape of a man : and the matter of a woman, the substance of a female plant ; with many other such dotish dreams. They fable further and affirm that he who would dig up a plant thereof must tie a dogge thereunto to pull it up, which will give a great shrike at the digging up ; otherwise if a man should do it, he should certainly die in short space after."

In the earliest herbals the plant is represented not only with the legs of a human being, but with the full body and head as well. It is so in the *Ortus Sanitatis*, A.D. 1476, where the man has a beard, and the woman long hair, both having the plant rising from their heads. For, in common with many other plants, this plant had two varieties—the male and the female. Plants with roots similar to the lower part of the body no doubt often occur, and such were hawked about by friars and other dealers in magical charms during the Middle Ages. The roots of the White Bryony were cut to serve as substitutes for Mandrakes, or were made to grow into moulds to take the required shapes. There are several allusions to the Mandrake's magical qualities in Shakespeare.

Blake, as we might expect, " worked in " the Mandrake.

The EGG-PLANT is very similar to the MAD-APPLE.

327

OLEACEÆ—OLIVE.

Trees or shrubs, with branches usually in pairs, and ending in a conspicuous bud. Leaves opposite, simple, sometimes pinnate. Flowers in axillary or terminal racemes or panicles—the pedicels opposite.

Calyx inferior, 4 cleft. Corolla regular, 4 cleft, sometimes of 4 separate petals. Stamens 2.

Style 1, or the stigma sessile on the ovary. Stigma bifid or undivided.

An Order which includes the Olive, Privet, Phillyrea, Ash, Lilac, and Manna Ash (Fraxinus Ornus).

The Lilac is Syringa, which must not be confused with the Mock-Orange or Syringa—Philadelphus— of the Order Philadelphaceæ.

The OLIVE is one of the slowest-growing and longest-living of trees. It seldom exceeds 30 ft. in height. Its leaves are evergreen, and like all evergreens become rather grey and dusty in their colour in hot climates. The trunks and branches are gnarled and straggling and ash-coloured. The leaves, particularly of the cultivated species, are hoary underneath. The wild species has spines : its fruit is worthless.

The ASH has flowers without either calyx or corolla. They have a bottle-shaped pistil with a stamen on either side. But some of the plants have flowers with stamens only. In the Linnæan System the Ash is in the section Polygamia Diœcia. The flowers are in opposite tufts a few inches from the ends of the twigs. They are succeeded by winged fruit, which, hanging on the tree all the winter, make noteworthy grey, or pale-orange, tufts in spring. The Ash is one of the latest trees to put out its leaves. Its twigs are more stick-like, and thicker, than those of most trees, and the buds are at first remarkably black, and later are a soft beautiful green. None of the figures in the herbals do justice to this tree. Nor does Fig. 639 do justice to the Lilac. The flowers should be multiplied tenfold. The three details have been added to the figure.

635.—OLIVE.
OLEA EUROPÆA. Z. 67.1.
White. 15 ft. June—Aug.

636.—WILD OLIVE.
OLEA OLEASTER. Z. 67.2.
White. 5 ft. June—Aug.

637.—PHILLYREA.
PHILLYREA MEDIA.　M. (1565) 172.
Flowers white or greenish-white.　Berries black.
Phillyreas are evergreen shrubs rising to 10 or 12 ft. in height, with a grey or light brown bark.
The leaves are opposite.

329

638.—PRIVET.
LIGUSTRUM VULGARE. Z. 267.
White. 8 ft. Fl., size as shown. June—Aug.

639.—LILAC.
SYRINGA VULGARIS. Z. 256.2.
Lilac, or white. 8 ft. Fl., $\frac{1}{2}$ in. May.

640.—ASH.
FRAXINUS EXCELSIOR. Z.191.
Corolla and calyx lacking. A tree. Ap.

JASMINACEÆ— JASMINE.

These differ very little from Oleaceæ, chiefly—in externals —in the calyx and corolla being in 5 to 8 divisions instead of in 4.

641.—YELLOW JASMINE.
JASMINUM FRUTICANS.
Yellow. 3 ft. Ap.—Oct.
Flowers appearing before the leaves, which are ternate and alternate, and suggest Broom.

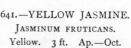

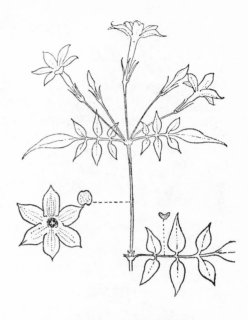

642.—WHITE JASMINE.
JASMINUM OFFICINALE. J. 892.1.
White. 15 ft. Fl., ⅝ in. June—Oct.

643.—WHITE JASMINE.
JASMINUM OFFICINALE.
The native country of this shrub is unknown.

644.—COWSLIP. [PRIMULACEÆ.]
PRIMULA VERIS. Z. 1134.1.
Deep yellow. 9 in. Fl., ⅝ in. May.

645.—OXLIP. [PRIMULACEÆ.]
PRIMULA ELATIOR. Z. 1134.2.
Pale yellow. 9 in. Fl., ¾ in. April.

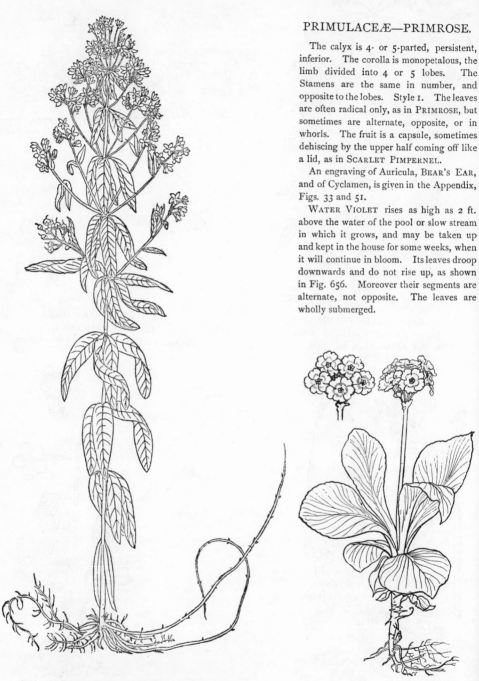

PRIMULACEÆ—PRIMROSE.

The calyx is 4- or 5-parted, persistent, inferior. The corolla is monopetalous, the limb divided into 4 or 5 lobes. The Stamens are the same in number, and opposite to the lobes. Style 1. The leaves are often radical only, as in PRIMROSE, but sometimes are alternate, opposite, or in whorls. The fruit is a capsule, sometimes dehiscing by the upper half coming off like a lid, as in SCARLET PIMPERNEL.

An engraving of Auricula, BEAR'S EAR, and of Cyclamen, is given in the Appendix, Figs. 33 and 51.

WATER VIOLET rises as high as 2 ft. above the water of the pool or slow stream in which it grows, and may be taken up and kept in the house for some weeks, when it will continue in bloom. Its leaves droop downwards and do not rise up, as shown in Fig. 656. Moreover their segments are alternate, not opposite. The leaves are wholly submerged.

646.—YELLOW LOOSESTRIFE. HERB PRYTHEE.
LYSIMACHIA VULGARIS. F. 492.
Yellow. 3 ft. Fl., ½ in. July.

647.—PRIMULA AURICULA.
Red, yellow, purple, shaded and varie-gated concentrically. 9 in. Fl., 1 in.

648.—PRIMROSE.
PRIMULA VULGARIS. J. 781.5.
Pale yellow. 8 in. Fl., 1 in. March—May.

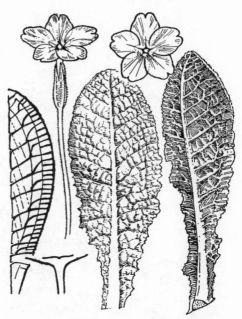

649.—DETAILS OF PRIMROSE.

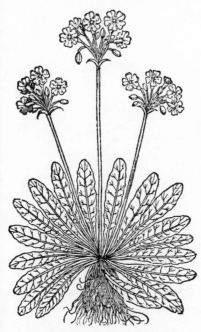

650.—BIRD'S-EYE PRIMROSE.
PRIMULA FARINOSA. G. 639.2.
Pale purple with a yellow eye. 6 in. June—July.

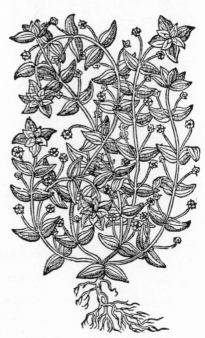

651.—SCARLET PIMPERNEL.
ANAGALLIS ARVENSIS. G. 494.1.

652.—BLUE-FLOWERED PIMPERNEL.

ANAGALLIS CÆRULEA. F. 19.

Leaves 5-nerved. Stem erect, a little winged. Cornfields. June—Sep. Rare in England.

This was called the female Pimpernel, the Scarlet Pimpernel was called the male.

334

653.—SCARLET PIMPERNEL. POOR MAN'S WEATHER-GLASS.

ANAGALLIS ARVENSIS. F. 18.

This figure is about half-scale. "It is remarkable with what a sparing hand nature has dealt out her richest and most glorious colour, for except this, and the poppies, I do not recollect any indigenous plant or a scarlet colour."—Dickenson.

Stem, 4-cornered ; leaves sessile, opposite. Petals scarlet. Anthers yellow. Leaves mottled with purple underneath, and 3-nerved. May—Aug.

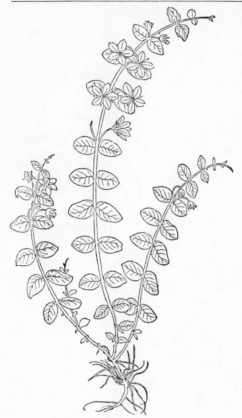

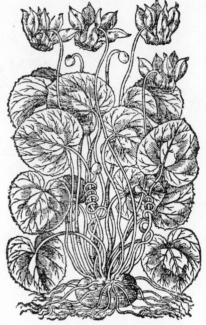

655.—CYCLAMEN. SOWBREAD.
CYCLAMEN EUROPÆUM. G. 694.1.
Pink. 8 in. Fl., 1 in. Aug.—Sep.

654.—MONEYWORT. HERB TWOPENCE.
LYSIMACHIA NUMMULARIA. F. 401.
Yellow. Fl., ¾ in. Trailing. May—July.

CYCLAMEN. — Although Fig. 655 makes a handsome
panel, the growth of the plant is more properly that given in
Fig. 33 of the Appendix, which is of the kind called the
IVY-LEAFED CYCLAMEN, Cyclamen hederifolium.   The
name, Cyclamen, refers to the numerous circles or coils into
which the flower-stalks twist when the flower has faded.
The leaves are splashed with white, in a concentric manner.

656.—WATER VIOLET.
HOTTONIA PALUSTRIS. C. E. 897.
Mauve, yellow centre. Fl., ¾ in. July.

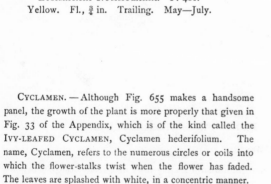

336

PLUMBAGINACEÆ—THRIFT.

Herbaceous plants or undershrubs, with alternate undivided leaves, without stipules. Flowers loosely panicled, or in heads. Calyx tubular, plaited, persistent, sometimes coloured—particularly in the Sea Lavender, wherein the calyxes appear to be the corollas. Corolla monopetalous, or of 5 petals. Stamens 5. Styles 5, 3, or 4.

657.—THRIFT.
ARMERIA MARITIMA.
Pink.　8 in.　Grows in a dense tuft.　June—Aug.

658.—SEA LAVENDER.
STATICE LIMONIUM.　M. 443.1.
Purplish-blue.　18 in.　July—Aug.

PLANTAGINACEÆ—PLANTAINS.

An Order of very lowly herbs possessing much beauty in the eyes of the designer on account of the parallel veining of the leaves, and of the rosettes which the leaves form.

The edges of the leaves in some species have slight teeth at wide intervals, and in Buckhorn Plantain have small lobes. The leaf-stalks are dilated at the base in those plants which have the larger leaves. In Buckhorn and Sea Plantain the leaves are narrow, fleshy and warty, with a strongly marked mid-rib.

Ribwort has its flower-stalk deeply ribbed and some 2 ft. in length.

The flowers are in spikes upon long stalks rising from the axils of the lower leaves. They appear, therefore, to come from behind the rosette.

The flowers are themselves in the axils of pointed bracts. The calyx is 4-parted—the parts well overlapping. It forms an egg-shape, and out of it shortly protrudes the long slightly feathery style. The corolla next appears exhibiting a 4-parted, star-like limb. Then follow the 4 stamens. The fruit is an egg-shaped capsule bearing at the top the remains of the style. The beauty of effect is entirely produced by the stamens—the spikes being otherwise green, and the corollas too inconspicuous, their colour being that of parchment.

For other figures of Plantains see the Appendix, Fig. 16.

337

659.—GREAT PLANTAIN.
PLANTAGO MAJOR. F. (1545).
Anthers yellow. 10 in. June—Aug.

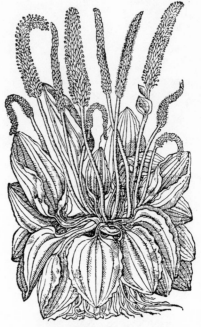

660.—GREAT PLANTAIN.
PLANTAGO MAJOR. G. 338.1.
Anthers yellow. 10 in. June—Aug.

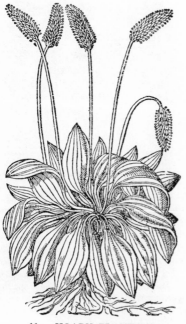

661.—HOARY PLANTAIN.
PLANTAGO MEDIA. G. 338.2.
Filaments violet. 10 in. June—Aug.

662.—RIBWORT PLANTAIN.
PLANTAGO LANCEOLATA. F. (1545).
Stalks ribbed. 1-2 ft. June.

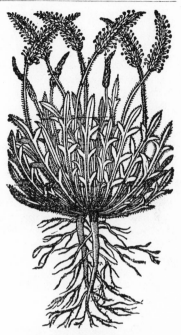

665.—HARTS-HORN.
PLANTAGO CORONOPUS. J. 427.1.
The plant in dry loose soil.

663.—BUCKHORN PLANTAIN.
PLANTAGO CORONOPUS. F. 1545.
Anthers yellow. 6 in. June—Aug.

664.—STAR OF THE EARTH.
PLANTAGO CORONOPUS.
The plant in dry short pasture, seen from above.

666.—NEVER-DYING FLEAWORT.
PLANTAGO CYNOPS. J. 587.2.
Shrubby. 6 in. May—Aug.

BORAGINACEÆ—BORAGE.

Herbs or shrubs ; their appearance remarkable for roughness and bristliness, for their simple leaves, and for their spiral inflorescence.

The distinguishing characteristics of the Order are—a naked style rising from amid the 4 distinct, seed-like, lobes of the ovary ; a monopetalous corolla, usually regular and 5- or 4-cleft, and often closed by scales at the throat ; 5 stamens, or 4, upon the corolla at the clefts ; a calyx usually very deeply 5-cleft, persistent ; inflorescence a one-sided spiral, spike, or raceme, or a panicle, or the flowers sometimes solitary and axillary ; bracts, if present, leafy, solitary, permanent ; leaves alternate for the most part, entire, sometimes wavy, or broadly toothed, often covered with bristles rising from hard bases ; stipules none ; stem rounded, with alternate axillary branches. Stems usually round.

The spiral flower-stalks are generally two together. These spiral peduncles gained some of the plants—Myosotis—the ancient name of Scorpion-grass.

Particularly notable is the change in the colour of the corolla in some cases, from pink-red in the bud to rich blue. Also is to be noticed " the white, or yellow throat of the flower, which is either completely, or imperfectly, closed with hollow convex valves ; or beset with swellings, or plaits, or dense hairs ; or entirely naked and pervious."—Smith.

The lobes of the ovary are so distinct that they have been called nuts, a term which if not correct is convenient. The calyx is persistent, and becomes a very conspicuous detail of the plant when the corolla has fallen. The spiral peduncle becomes straight as it uncoils.

These plants are mucilaginous, and mostly harmless. The roots of Alkanet of certain species yield red colouring or dyeing matter. Common Borage " gives a coolness to beverages in which its leaves are steeped."—Lindley.

The HELIOTROPES are sometimes separated from the Boraginaceæ and classed in an Order Ehretiaceæ.

A common plant of cultivation is CHERRY PIE, Heliotropium peruvianum.

667.—COMMON BUGLOSS.　　　　　　　　668.—"SMALL TURNSOLE."
LCOPSIS ARVENSIS. F. 269.　　　　　　　　　　G. 264.2.
Blue. 1 ft. Fl., ⅔ in. June.

669.—COMMON COMFREY.

SYMPHYTUM OFFICINALE.   M. (1565) 961.

2–3 ft.   Fl., ¾ in. long.

Yellow, purple, rose or white.   Stems winged as shown.   Watery places, river sides.

May—September.

COMFREY is reputed a wound-wort, and therefore anciently was called " Consolida."   Bentley says the root is excellent for bandages for broken limbs.

341

670.—COMMON COMFREY.
SYMPHYTUM OFFICINALE. F. 695.

671.—COMMON ALKANET.
Anchusa officinalis.　M. 447.2.
Purple.　2 ft.　Fl., $\frac{2}{3}$ in.　June—July.

672.—VIPER'S BUGLOSS.
Echium vulgare.　M. 448.
Blue.　2 ft.　Fl., $\frac{1}{2}$ in.　June—Aug.

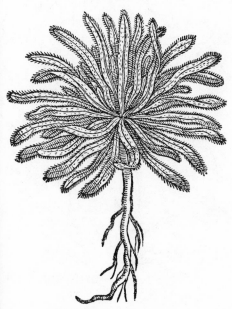

673.—VIPER'S BUGLOSS.
Echium vulgare.　M. 412.2.

674.—COMMON BORAGE.
Borago officinalis.　M. 513.1.

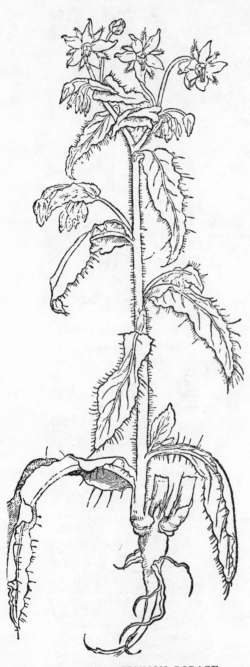

675.—COMMON BORAGE.
BORAGO OFFICINALIS.
Light blue. 18 in. Fl., 1 in. Brunf. 113.
June—Aug.

676.—OX-TONGUE.
Brunf.

344

677.—ANCHUSA PANICULATA. M. 573.2.
Blue. 2 ft. May—June.

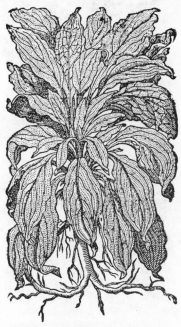

678.—HOUND'S-TONGUE.
CYNOGLOSSUM OFFICINALE. D. 1263.1.

679.—HOUND'S-TONGUE.
CYNOGLOSSUM OFFICINALE. G. 659.
Purplish-red. 2 ft. July.

680.—HOUND'S-TONGUE.
CYNOGLOSSUM OFFICINALE. D. 1262.1.

681.—BROAD-LEAFED LUNGWORT
PULMONARIA OFFICINALIS. G. 662.1.
Purple. 1 ft. Fl., ⅔ in.
May.

**682.—NARROW-LEAFED COWSLIPS
OF JERUSALEM.**
PULMONARIA OFFICINALIS. G. 663.3.
Leaves spotted with white.

683.—FORGET-ME-NOT.
MYOSOTIS PALUSTRIS.
Pale blue, yellow eye. 1 ft. Fl., ½ in.
June—Aug. Br.

684.—EVERGREEN ALKANET.
ANCHUSA SEMPERVIRENS. G. 653.3.
Deep blue. 1 ft. Fl., ⅝ in.
May—June. Br.

685.—RED ALKANET.
ANCHUSA. J. 800.1.

686.—GROMWELL.
LITHOSPERMUM OCHROLEUCUM. M. 420.2.
Pale yellow. 18 in. June.

687.—CREEPING GROMWELL.
LITHOSPERMUM PURPURO-CÆRULEUM. L. i. 458.1.
Purple-blue. 18 in. June—Aug.

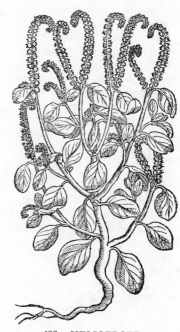

688.—HELIOTROPE.
HELIOTROPIUM EUROPÆUM. M. 561.2.
White. 9 in. June—Oct.

LABIATÆ—SAGE.

The plants of this Order are readily recognized ; though there are plants in other Orders, notably in the Scrophulariaceæ, which might be taken to belong to them. The first peculiarity is the form of the flower. The calyx is either regular and 5-toothed, or 2-lipped and irregular. The corolla is monopetalous, 2-lipped. The lower lip is generally divided into 3 lobes, with the central one often again divided into 2 ; and the upper lip is very often hood-like, but sometimes is small, as if it had been cut off. Corollas of a similar form will be seen in the Scrophulariaceæ and in Acanthus. The stamens are 2 or 4 ; if 4, then 2 are much larger than the others. Style 1 ; stigma forked. The flowers are in the axils of the upper leaves, or of bracts. These bracts are a notable feature of the plants, and are sometimes very large, especially in CLARY, Fig. 693, and in Prunella (see Appendix, Fig. 54), which is a common flower in meadows, with a blackish cylindrical head, bearing bright purple flowers, a few at a time. This cylindrical head consists very largely of bracts.

689.—SAGE FLOWERS.

Other peculiarities of these plants are the square stems and the opposite heart-shaped leaves. Some Scrophulariaceæ, however, have square stems and opposite simple leaves. The ultimate test when the plants otherwise agree is in the form of the ovary. In Labiatæ there are, as it were, 4 little round nuts clustered round the base of the style, whereas in Scrophulariaceæ the ovary is pear-shaped, with the style atop of it.

This difference in the ovary provided a distinction in the Linnæan System by which the Didynamia were divided into Gymnospermia, or naked-seeded, and Angiospermia, in which the seeds were enclosed in a seed vessel. The former are Labiatæ, the latter Scrophulariaceæ.

Another difference between the two Orders is in the useful, harmless, and aromatic nature of the Labiatæ, and the rather poisonous and useless character of the others.

The leaves of Labiatæ are serrated, and have no stipules. They generally possess receptacles of aromatic oil, and are somewhat granular in texture.

The flowers are single or in cymes, in the axils of the leaves, or bracts, and make a whorled effect. Being generally small in size, they have not in many cases been rendered with much truth of detail in the old wood-cuts. An exception is certainly present in the drawing of the Dead Nettle, from Brunfels. Fig. 710.

690.—THREE-LOBED SAGE.
SALVIA TRILOBA. M. 338.2.
Red. 2 ft. June—July.

691.—GARDEN SAGE.

SALVIA OFFICINALIS.   F. 248.

Stem round.   Whorls few—flowered, crimson-red.   2 ft.   June—July.

692.—MEADOW-SAGE OR CLARY.

SALVIA PRATENSIS. F. 569.

Purple. 3 ft. Fl., 1 in. long. Corolla, thrice as long as the calyx. June.

693.—COMMON CLARY.

Salvia Sclarea.　F. 568

The bracts are coloured and are longer than the calyx.　The corolla is light blue.　4 ft.　July—Sep.

ANNUAL CLARY. The bracts are pinkish, the upper ones large and not enclosing flowers. The flowers are white with the upper lip brown. The leaves (according to Parkinson) are broad, rough, wrinkled and whitish. Clary means clear-eye, from the use of the powdered seeds, mixed with honey, as a salve for the eyes.

FRENCH SAGE. Gerarde says the flowers are yellow and much larger than those of the Dead Nettle. There are plants of this Order with large flowers in terminal whorls, as in this case, in cultivation in greenhouses. The term " French " merely meant foreign.

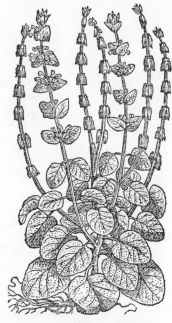

694.—ANNUAL CLARY.
SALVIA HORMINUM. M. 411.

695.—"GREAT CAT-MINT." G. 554.2.

696.—"FRENCH SAGE."
"VERBASCUM SYLVESTRE." M. 499.4.

697.—CAT-MINT.
NEPETA CATARIA.　F. 434.

Flowers pale pink, with purple spots.　Leaves downy.　2–3 ft.　Hedge-banks.　July—Sep.

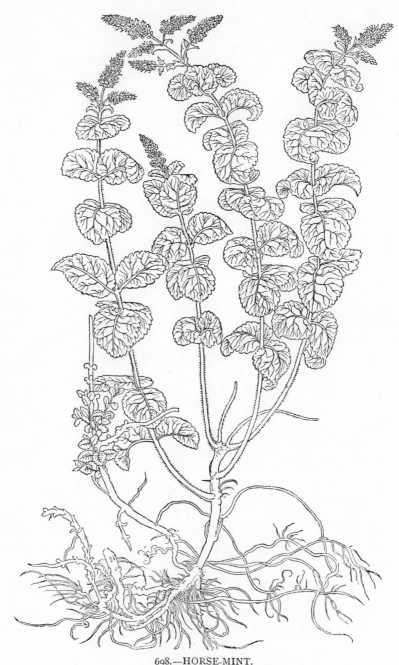

698.—HORSE-MINT.
MENTHA SYLVESTRIS.　F. 289.

Flowers pale lilac.　Bracts awl-shaped.　2–4 ft.　Moist ground.　Aug.—Sep.

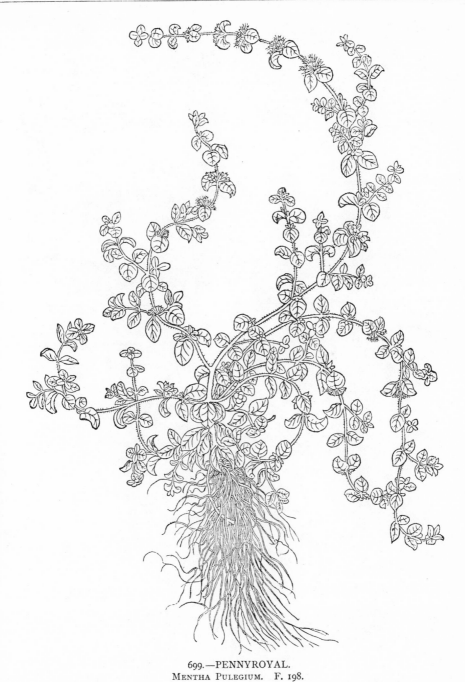

699.—PENNYROYAL.
Mentha Pulegium. F. 198.

Flowers pale purple. Stems prostrate. A naturalized plant growing in wet places. Height 8 in. Aug.—Sep.

700.—COMMON CALAMINT.
CALAMINTHA OFFICINALIS, OR THYMUS CALAMINTHA.   M. (1565) 716.
Flowers pale purple, in one-sided cymes.   Height 18 in.   Waysides.   July—Sep.

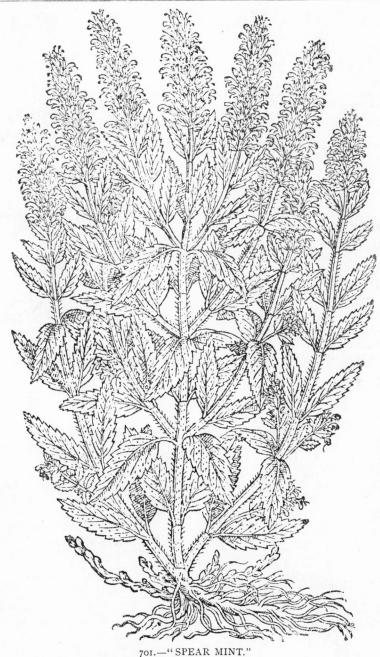

701.—"SPEAR MINT."
MENTHA VIRIDIS (?). G. 553.2.

SPEAR MINT is the common culinary herb—Mentha viridis—but this figure has been passed over by botanists who have referred to these old illustrations—possibly because in Spear Mint the spike of flowers is interrupted in the same manner as are the spikes in the next three figures. It may be a continental species. SPEAR MINT is 18 inches high, and bears reddish-purple flowers in August and September. This figure is enlarged from 4¾ inches high, and, although it has not reproduced well, shows the high merit of the engraving.

357

702.—WATER OR HAIRY MINT.
MENTHA AQUATICA OR HIRSUTA. D. 677.1.
So these have been identified, but the spikes are not dense enough.

703.—WATER MINT.
MENTHA AQUATICA. D. 674.2.
? Horse Mint. (See Fig. 715.)

704.—LESSER CALAMINT.
CALAMINTHA NEPETA. M. 341.1.
Purple. 1 ft. Aug.—Sep.

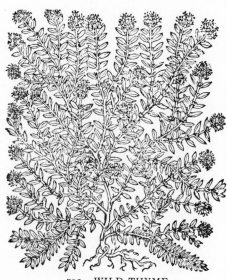

705.—WILD THYME.
THYMUS SERPYLLUM. G. 455.1.
Reddish. 2 in. June—Sep.

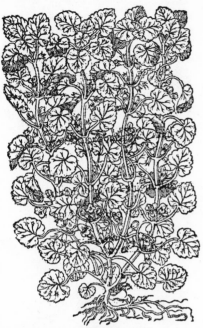

706.—GROUND IVY.
NEPETA GLECHOMA.　G. 705.
Blue or white.　4–9 in.　April—Oct.

707.—WOOD BETONY.
BETONICA OFFICINALIS.　M. 430.
Purple-pink.　1–2 ft.　July—Aug.

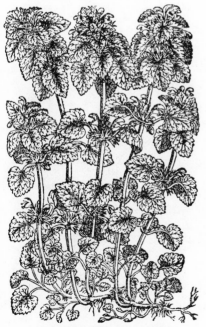

708.—RED DEAD NETTLE.
LAMIUM PURPUREUM.　G. 568.4.
Purple or pink.　6–10 in.　All the year.

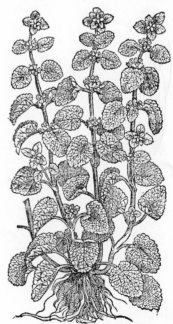

709.—STINKING HOREHOUND.
BALLOTA NIGRA.　M. 390.
Purple.　2–3 ft.　July—Aug.

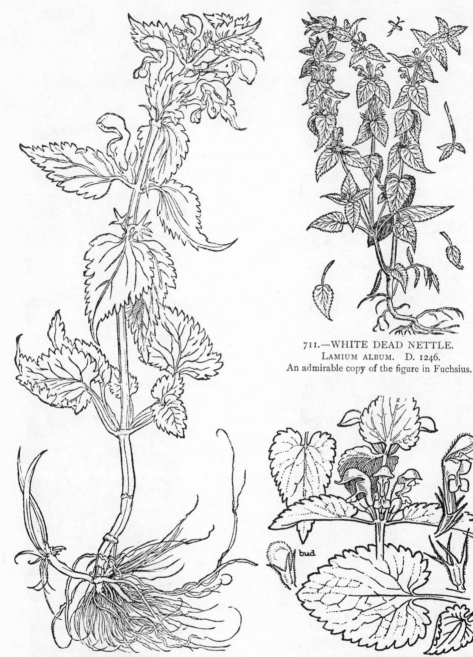

711.—WHITE DEAD NETTLE.
LAMIUM ALBUM. D. 1246.
An admirable copy of the figure in Fuchsius.

710.—DEAD NETTLE. B. 153.
Brunfels gives two very similar figures. The other is
white, and this is coloured red, but it accords to the form
of the White not of the Red Dead Nettle.

712.—WHITE DEAD NETTLE.
LAMIUM ALBUM.
White, anthers black. 18 in. April—Oct.

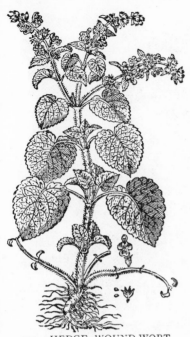

713.—HEDGE WOUND-WORT.
STACHYS SYLVATICA. J. 704.5.

714.—WOOD-SAGE.
TEUCRIUM SCORODONIA. J. 662.
Pale yellow. 1–2 ft. July—Sep.
The flower has the upper lip cut away
and the lower lip long and pendulous.

715.—WATER MINT.
MENTHA HIRSUTA.
Pale purple. 1–2 ft. Aug.—Sep.

Pink-Lilac
Crimson
Green.

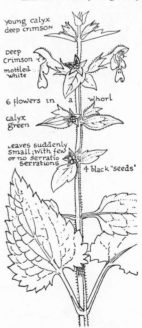

young calyx
deep crimson

Deep
Crimson
mottled
white

6 flowers in a whorl

calyx
green

Leaves suddenly
small; with few
or no serratio
Serrations

4 black "seeds"

716.—HEDGE WOUND-WORT.
STACHYS SYLVATICA.
2–3 ft. July—Aug.

717.—CUT-LEAFED GERMANDER.
Teucrium Botrys. M. (1565) 819.
Flowers red. Height 9 in. South Europe. July—Sep.

HERB TEUCER.
718.—TEUCRIUM FLAVUM. F. 829.
Flowers yellow. Height 2 ft. South Europe. June—July.
Said by Pliny to have been first used medicinally by Teucer, the Trojan prince.

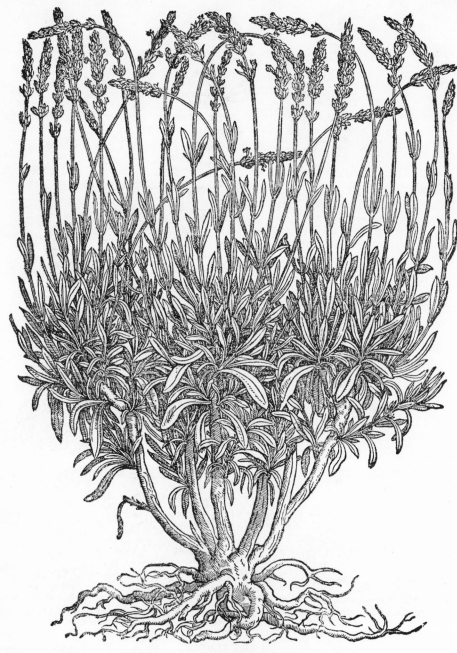

719.—LAVENDER.

LAVANDULA SPICA. M. (1565) 31.

Flowers lilac. Height 2 ft. South Europe. Gardens. July—Sep.

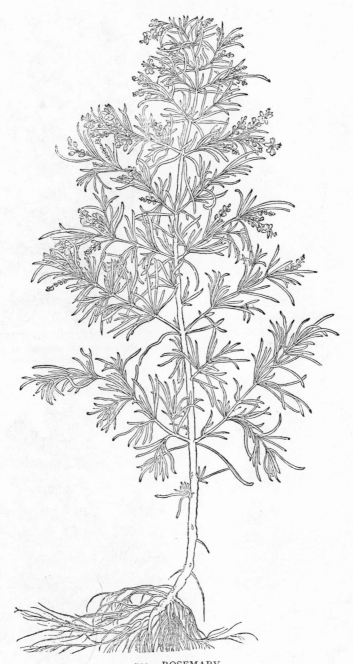

720.—ROSEMARY.
ROSMARINUS OFFICINALIS. F. 478.
Flowers purple. Height 4 ft. South Europe. Jan.—April. Gardens.
The inflorescence is incorrect in this beautiful figure.

721.—WINTER SWEET MARJORAM.
ORIGANUM HERACLEONTICUM. M. 334.2.
White. 1 ft. June—Nov.

722.—DITTANY OF CRETE.
ORIGANUM DICTAMNUS. M. 337.1.
Pink. 1 ft. June—Aug.

723.—WATER HOREHOUND.
LYCOPUS EUROPÆUS. M. 451.1.
White. 1–2 ft. June—Sep.

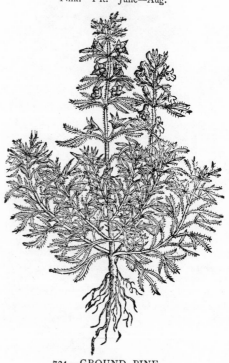

724.—GROUND PINE.
AJUGA CHAMÆPITYS. L. i. 385.1.
Yellow. 6 in. April—July.

725.—POLY.
TEUCRIUM POLIUM. M. 396.1.
Plant hoary. 1 ft. July—Sep.

726.—GOLDEN POLY.
TEUCRIUM FLAVESCENS. M. 396.2.
Yellow. 1 ft. July—Sep.

HYSSOP.—Parkinson says it "is a small bushy plant, not rising above 2 feet high, with many branches, woody below, and tender above, whereon are set at certain distances, sundry small long and narrow green leaves : at the top of every stalk stand bluish-purple gaping flowers, one above another in a long spike or ear. . . . The whole plant is of a strong sweet scent." In the flower garden he mentions the White, the Russet, the Yellow or Golden, and the Double kinds. The last he says is suitable for making borderings to knots of flowers, as it will not grow too woody or great. The others have variegated leaves, green-spotted or splashed with white, ash-colour, and yellow, and produce such a pleasant effect that they "provoke many gentlewomen to wear them in their heads and on their arms."—*Paradisus*, p. 455.

No satisfactory conclusion seems to have been reached as to what the Hyssop of Scripture was—whether the Hyssopus officinalis described by Parkinson, or Marjoram, or the Caper. The inflorescence of Hyssop differs from that of Poly in being a spire, or, may we say, a fox's brush ; otherwise in form the plants are not unlike. The name Poly refers to its white or hoary appearance.

A plant of a very different character is Hedge Hyssop of the neighbouring Order, the Scrophulariaceæ. It was long used as a cathartic.

VERBENACEÆ—VERVAIN.

COMMON VERVAIN is the only European member of this Order, which differs from Labiatæ in that the 4 nuts of its fruit are consolidated together and the style is terminal, and not, as in Labiatæ, rising up between the nuts from their base. The calyx is tubular inferior, persistent; the corolla tubular, with usually an irregular limb, more or less 2-lipped. Stamens 4, usually didynamous, sometimes 2. Leaves simple or compound, opposite or alternate, without stipules. Herbs, shrubs or trees.

Many exotics are showy garden plants; others produce edible fruits, and others valuable timber.

Vervain has the honour of being anciently in great repute; it was used in incantations, sacrificial rites (though *verbena* was the word used for a sacrificial herb, whatever it was). See page 4 for an early representation of the plant.

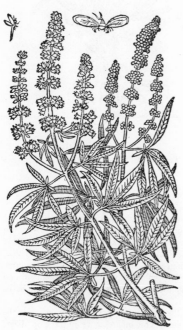

727.—CHASTE TREE.
VITEX AGNUS-CASTUS. D. 281.1.
Blue or white. 6 ft. Autumn.

728.—BROAD-LEAFED CHASTE TREE.
VITEX AGNUS-CASTUS. D. 281.2.
The seeds anciently considered anti-aphrodisiac.

729.—COMMON VERVAIN.
VERBENA OFFICINALIS.　M. (1565) 1052.
Stem erect, 1–2 ft.　Flowers lilac, in slender panicled spikes.　Waste places.　July.

ACANTHACEÆ—ACANTHUS.

The ACANTHACEÆ correspond in several particulars to the Labiatæ and Scrophulariaceæ, near which they are placed. The stamens are 2, or 4, in which case they are didynamous. The calyx and corolla are 2-lipped. The leaves are opposite The fruit is a capsule, shown by Fuchsius in a separate detail.

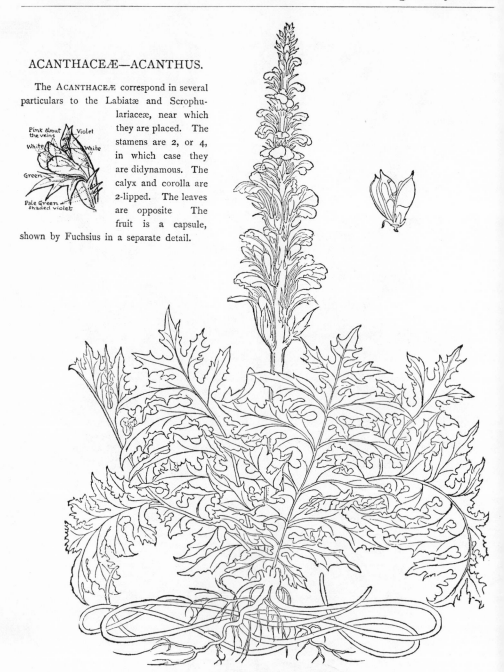

730.—PRICKLY ACANTHUS.
ACANTHUS SPINOSUS. F. 52.
Height 3 ft. The flowers purple and white, each in the axil of a bract.
Every one knows the story of the origin of the Corinthian capital—not very convincing
to a designer. The plant is native in South Europe. July—Sep.

SCROPHULARIACEÆ—TOAD-FLAX.

Herbs, or sometimes shrubs, usually scentless, sometimes fœtid, rarely aromatic, similar in certain outward respects to Labiates. This similarity is due to the shape of the leaves, to their being opposite, to the lipped character frequently to be seen in the flowers, to their growing in the axils of the uppermost leaves somewhat in the same way, and to their having occasionally 4-cornered stems.

However, they differ considerably from Labiates, and many of them bear very little resemblance to those plants. Such are the Mulleins and Veronicas.

The Order is named from Scrophularia, Figwort. This has an ovoid corolla, 5-lipped. The ovoid shape is hardly elegant, but rather of a corpulent outline, and the axis is bent. It is one of the plants with a square stem.

The fruit in this Order is somewhat pear-shaped, with a bent axis, and is either broader below or above. Its two cells are evident in its form.

The calyx is 4- or 5-parted, and sometimes is distinctly 2-sided, as in Mimulus and Yellow Rattle, having, as it were, a keel above and below. Sometimes, however, the 5 parts are almost so many sepals, as in Foxglove.

The corolla is tubular with a curved axis, as if the mouth of the tube were directed upwards. At the mouth, the limb divides into 5 or 4 parts. These parts are obscurely marked in Foxglove, but in Mullein they appear like 5 petals, and in Veronica like 4. The petal-like segments may be classed as 2 above and 3 below, which in some plants as Snap-dragon and Toad-flax are positively mouth-like.

The tube of the corolla is sometimes spurred below, as in Toad-flax. In some plants the corolla is so much bent that, being spurred also, it looks as if it were sitting in, rather than arising from, the calyx.

The stamens are 2 or 4. If 4, then 2 are larger than the others. Sometimes there is a fifth or a rudimentary fifth. Style 1, sometimes shortly bifid.

The fruit sometimes looks like the skull of an animal, 2 holes in it being like eyes.

The leaves are opposite, whorled, or alternate. The veins do not proceed to the points of the teeth (Fig. 753). Lindley says the species are "generally acrid, bitterish and suspected."

A Shrub Veronica is common in gardens. For figures of Foxglove and Veronica see Appendix, Fig. 54.

In colour the corollas are of great variety, and are generally variegated. The MULLEINS are purely yellow; but usually one or both lips differ from the rest of the corolla. Figwort is green tingeing into purple; Water Betony is green with deep purple lips; Foxglove is pink-purple with a pale or white portion mottled with brownish-purple inside: it is also yellow towards the base; Snapdragon has one or both lips decidedly different from the remainder; and Veronica, which is of a beautiful blue, is generally partially white, or perhaps with a tinge of red.

731.—WHITE MULLEIN.
VERBASCUM RHOMBIFOLIUM. F. 847

732.—GREAT MULLEIN.　HIG TAPER.
Verbascum Thapsus.　F. 846.
Yellow.　Height 6 ft.　Flowers 1½ in.

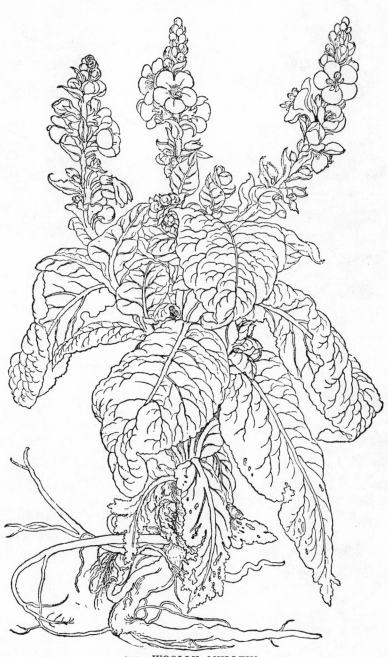

733.—WOOLLY MULLEIN.
VERBASCUM PHLOMOIDES. F. 848.
Yellow. Height 3 ft. Italy. July—August.

734.—WHITE MOTH MULLEIN.
VERBASCUM BLATTARIA. L. i. 563.1.
Yellow. 3–4 ft. Fl., ¾ in. July—Nov.

735.—DARK MULLEIN.
VERBASCUM NIGRUM. G. 631.2.
Yellow. 3 ft. Fl., ¾ in. June—Oct.

736.—GREAT SNAPDRAGON.
ANTIRRHINUM MAJUS. L. i. 405.1.
Purple, pink, or white. 1–2 ft. Fl., 1½ in.
June—Sep.

737.—LESSER SNAPDRAGON.
ANTIRRHINUM ORONTIUM. Z. 1160.3.
Pink purple, lip white. 9 in. Fl., ⅝ in.
July—Aug.

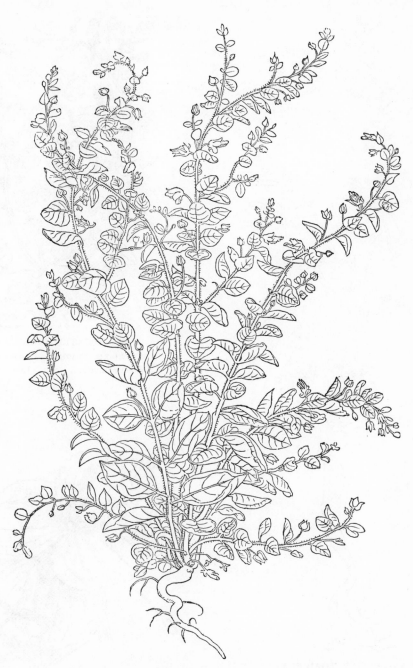

738.—ROUND-LEAFED TOAD-FLAX.   FLUELLEN.
LINARIA SPURIA.   F. 167.
Violet and yellow.   Leaves downy.   2-4 in.   Fl., ⅜ in.   July—Sep.

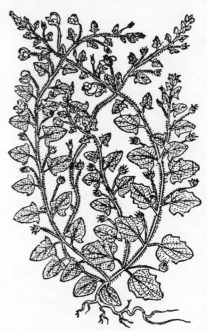

740.—POINTED LINARIA.
LINARIA ELATINE. G. 501.2.
Violet and yellow. 4 in. June—Sep.

739.—YELLOW TOAD-FLAX.
LINARIA VULGARIS. F. 545.
Bright yellow, orange lip. 1–2 ft. Fl., 1 in.
June—July

741.—MIMULUS.
MIMULUS LUTEUS.
Yellow. 2 ft. Fl., 1 in. June—Sep.

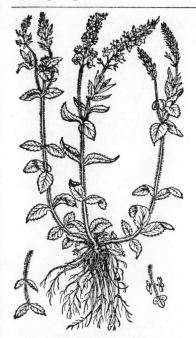

742.—VERONICA TEUCRIUM.　D. 1162.2.
Light blue.　**2** ft.　June—Aug.

743.—BROOK-LIME.
VERONICA BECCABUNGA.　G. 496.4.
Blue.　6–8 in.　Fl., ¼ in.　June—July.

744.—IVY-LEAFED SPEEDWELL.
VERONICA HEDERIFOLIA.　G. 493.3.
Pale blue.　2–5 in.　Fl., ¼ in.　April—Aug.

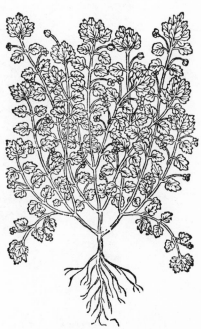

745.—GERMANDER CHICKWEED.
VERONICA AGRESTIS.　G. 492.1.
Pale blue.　2–6 in.　Fl., ¼ in.　April—Sep.

746.—COMMON SPEEDWELL.
VERONICA OFFICINALIS.　D. 1319.2.
Blue.　4–6 in.　Fl., $\frac{3}{8}$ in.　May—Aug.

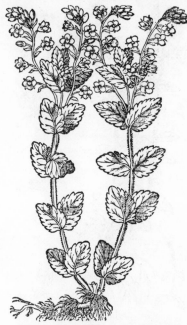

747.—GERMANDER SPEEDWELL.
VERONICA CHAMÆDRYS.　G. 530.4.
Bright blue.　4 in.　Fl., $\frac{1}{2}$ in.　May—June.

748.—DWARF RED RATTLE.
PEDICULARIS SYLVATICA.　L. 748.2.
Pink or white.　3–8 in.　Fl., 1 in.　June—July.

749.—WATER BETONY.
SCROPHULARIA AQUATICA.　J. 715.
Greenish, lips purple.　3–4 ft.　Stem
square.　July.

750.—KNOTTED FIGWORT.
SCROPHULARIA NODOSA. M. (1565) 1130.
Greenish purple. 2–3 ft. Stem square. July.

751.—PURPLE COW-WHEAT.
MELAMPYRUM ARVENSE. J. 90.3.
Yellow and purple. 18 in. Fl., 1 in. July.

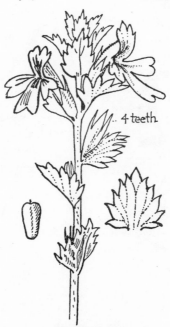

752.—RED EYEBRIGHT.
BARTSIA ODONTITES.
Red-purple. 9 in. July—Sep.

753.—EYEBRIGHT.
EUPHRASIA OFFICINALIS.
White or pinkish. 8 in. Fl., ½ in.
long. July—Sep.

754.—YELLOW RATTLE.
RHINANTHUS CRISTA-GALLI.
Yellow. 1 ft. Fl., ⅝ in. long. June.

[Here end the CoROLLIFLORÆ, and here begin the MONOCHLAMYDEÆ, which have not both a calyx and corolla, but one of them only, which is sometimes called a calyx, but generally a perianth—it is sometimes corolla-like, as in Begonia.]

POLYGONACEÆ—BUCKWHEAT. DOCKS.

Herbs with alternate leaves usually with membranous stipules, which cohere round the stem. Leaves rolled back when young. Flowers usually in racemes or spikes. Calyx often coloured. Corolla o. Fruit, a triangular nut.

The word "Polygonaceæ" refers to the many knees or angles which the stems present.

The Rhubarb of the kitchen garden is Rheum Rhaponticum. The drug Rhubarb is the dried root of Rheum palmatum. Both are Asiatic plants. Among European plants are some which have the same medicinal value— Monk's Rhubarb and Herb Patience, both of which are illustrated in the figures.

The Order presents many plants which have value as food. Some have been used as pot-herbs and for salads, particularly Common Sorrel, Rumex Acetosa, and Herb Patience, Rumex Patientia.

The fruit of Common Buckwheat, Fagopyrum esculentum, is used as a substitute for wheat.

755.—BUCKWHEAT.
POLYGONUM FAGOPYRUM. G. 82.2.
Pink. 1 ft. July—Sep.

756.—SPOTTED PERSICARIA.
POLYGONUM PERSICARIA.
Pink. 1–2 ft. July—Oct.

757.—BLACK BINDWEED.
POLYGONUM CONVOLVULUS.
Greenish. 4–6 ft. June—Sep.

758.—SMALL SHEEP'S SORREL.
RUMEX ACETOSELLA. G. 321.6.
Red or yellowish. 6–12 in. June—Aug.

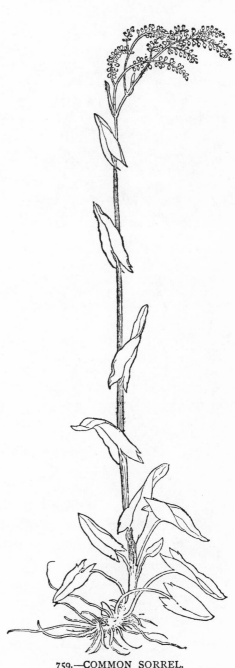

759.—COMMON SORREL.
RUMEX ACETOSA. F. 464.
Red. 1–2 ft. May—July.

760.—SHEEP'S SORREL.
RUMEX ACETOSELLA.　J. 397.3.
Red or yellowish.　6–12 in.　June—Aug.

761.—GRAINLESS WATER DOCK.
RUMEX AQUATICUS.　D. 606.1.
Green.　2–3 ft.　July.

762.—WATER DOCK.
RUMEX HYDROLAPATHUM.　L. i. 285.2.
Green.　3–5 ft.　July—Aug.

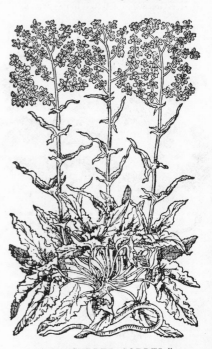

763.—"CURLED SORREL."
"OXALIS CRISPA."　G. 320.5.
"A stranger in England."

764.—GRAINLESS WATER DOCK.
RUMEX AQUATICUS. M. (1565) 449.
Flowers green. Height 2–3 ft. July.
(Fig. 761, which represents the same plant, must be regarded as decorative rather than botanical.)

765.—BISTORT OR SNAKE-WEED.
POLYGONUM BISTORTA. F. 774.
Stem quite simple, with a single spike.   Leaves egg-shaped, running down the leaf-stalks.   Upper
leaves arising from long sheaths enclosing the stem.
Flower spike, pink.   Height 18 in.   May—June.

766.—MONK'S RHUBARB.
RUMEX PATIENTIA. F. 462.
Monk's Rhubarb has somewhat of the medicinal value of Turkey Rhubarb, and was substituted for it.
France. 1 ft. high.

CHENOPODIACEÆ—GOOSE-FOOTS.

The chief plants of this Order are GOOSE-FOOT, ORACHE, BEET, MANGOLD-WURZEL and SPINACH. Their names indicate their general use as food. Some of them, and particularly Salicornia and Salsola, are burned to produce carbonate of soda. The goose-foot-like leaf will readily be recognized in the illustrations. The leaves are generally alternate and are without stipules. From their axils rise spikes of flowers, sometimes with, sometimes without, small leaves upon them. The flowers are green, minute, with a calyx generally of 5 clefts, 5 stamens and 1 style with 2 stigmas. But there is considerable variation in the details of the flower, especially as the sexes are sometimes divided. Fruit 1-seeded, enclosed in the perianth.

The leaves are often fleshy in substance, and exhibit a mealy or frosted appearance. They are often strongly tinged with pink, red or purple—the colour-effect being very beautiful. In Salsola (Saltwort) the leaves are awl-shaped, in Salicornia (Glasswort) there are none.

The common habitat of the Chenopods is in salt marshes.

While some few are used in medicine, Bentley does not refer to any having dangerous qualities.

Cocks-comb, Prince's Feather and Love-lies-bleeding are garden plants in the nearly allied Order AMA-RANTHACEÆ. The spikes of flowers, which in Docks and Chenopods are not showy, become in these dense and crimson.

767.—WALL GOOSE-FOOT.
CHENOPODIUM MURALE. G. 256.2.
Green. 1 ft. Aug.—Sep.

768.—GARDEN ORACHE.
ATRIPLEX HORTENSIS. G. 256.1.
Green. 6 ft. July—Aug.

769.—SPINACH.
SPINACIA OLERACEA. D. 544.
Green. 18 in. Mar.—Oct.

770.—FROSTED SEA ORACHE.
ATRIPLEX LACINIATA. Z. 554.2.
Greenish-yellow. 8 in. July.

Frosted is the best description of this plant, which has its pale green colour tinged pink and purple, and covered all over with a white meal-like frost. Seashores.

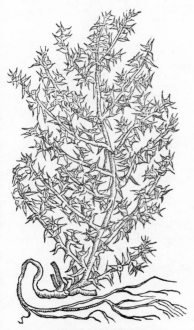

771.—PRICKLY SALTWORT.
SALSOLA KALI. M. 461.
Pinkish. 1 ft. July.

GLASSWORT, Salicornia. Its juicy and jointed branches have the appearance of having been stripped of their leaves. Grows in salt marshes and muddy shores. Height about 1 foot.

772.—RED GOOSE-FOOT.

CHENOPODIUM RUBRUM. F. 653.

Varies in colour from pale green to a fine purple, and in height, according to the poorness or richness of the soil, from being procumbent and dwarfish, to being erect and as tall as 2 or 3 feet. Rubbish. Aug.

773.—GOOD KING HENRY. ENGLISH MERCURY.
CHENOPODIUM BONUS-HENRICUS. F. 463.
Height, 1 foot. Cultivated sometimes as a perennial spinach. Road-sides. Rubbish. May.

774.—RED BEET.
BETA VULGARIS. F. 213.
Height, 4 ft. Stems and veins of the leaves red. S. Europe. Aug.

SEA BEET, Beta maritima, 2 feet high, is native on English seashores. Its leaves are more angular than these. Stokes identified this plant as TURNIP-ROOTED BEET, Beta esculenta raposa.

775.—COMMON GREEN BEET.
BETA ESCULENTA VIRIDIS. F. 806.

Usually called Beta nigra, black (red) beet, in the herbals, but Dalechampius called it Beta candida, white beet, which is Beta Cicla, a favourite vegetable in South Germany. Rises to 6 feet high. Aug.

The name given above is that by which Stokes identified it. The flowers are in fours, threes, and twos. The plant is often purplish.

THYMELACEÆ.

Shrubs with entire leaves without stipules. Flowers with a coloured calyx, tubular, with 4 or 5 lobes—the shape very well expressed in this figure. The flowers come out before the leaves, and are followed by berries, which are red in MEZEREON, which is also called SPURGE OLIVE or DWARF BAY, and black in SPURGE LAUREL, Daphne Laureola, which rises to 8 feet and has greenish-yellow flowers in axillary racemes. Those of MEZEREON are 3 together.

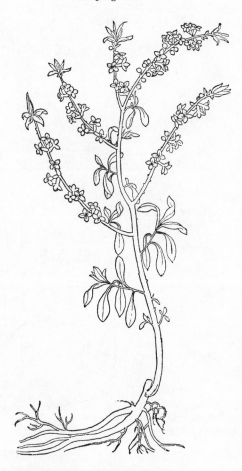

776.—OAK OF JERUSALEM.
CHENOPODIUM BOTRYS. M. 401.2.
Green. 1 ft. June—Sep.

Gerarde says that the leaves being cut like those of the oak, caused "our women" to call it Oak of Jerusalem, and Oak of Paradise. The reason does not seem clear, unless its properties as a destroyer of moths can have gained it such high distinction. Native of South Europe. It has also medicinal qualities to recommend it.

777.—MEZEREON.
DAPHNE MEZEREON. F. 227.
Pink or white. 4–8 ft. Fl., ½ in. Berries red.
March.

ELÆAGNACEÆ.

An Order not greatly differing from Thymelaceæ and Lauraceæ. SEA BUCKTHORN or SALLOW THORN is a British plant with silvery-white leaves, and the branches ending in thorns. It is a deciduous shrub.

LAURACEÆ.

An Order including CINNAMON and CAMPHOR, but chiefly interesting to artists on account of the

Sweet Bay Tree, or True Laurel. The Laurel is a diœcious plant, the stamens being on one plant and the pistils on another. Details of the male flowers are given. The calyx is deeply 4-cleft. The colour is that of Meadow-Sweet.

778.—SEA BUCKTHORN.
HIPPOPHAË RHAMNOIDES. C. E. 81.
Berries orange. 4–10 ft. May.

779.—LAUREL.
LAURUS NOBILIS.
Yellow-white. 15 ft. April—May.

780.—LAUREL. BAY TREE.
LAURUS NOBILIS. M. (1565) 131.

Leaves deep green, inclining to olive. Berries dark purple or black when ripe. The flowers, leaves, and berries have all a fragrant odour. Native of South Europe. "In Southern Italy it grows to sufficient height to be considered a tree, and is so prolific in suckers and low shoots as always to have the character of a shrub."—Loudon. The emblem of Victory and Achievement.

ULMACEÆ—ELM TREES.

Handsome trees with rough alternate leaves with deciduous stipules.

The flowers come out before the leaves and handsomely stud the twigs in large purplish bunches. A detail from Lindley of Ulmus campestris is here given. The young twigs of leaves have at the base of each leaf a pink scale. These scales at first form the leaf-bud, see at the left side of Fig. 781.

In the WYCH or SCOTCH ELM the central twig always dies—the branches therefore constantly divide, and the great boughs and trunk, or trunks, are due to the survival of the fittest. Few trees are more beautiful in line.

The leaves vary in the different species in the measure of equality between the two sides of the leaf. They are very unequal in the SMOOTH ELM, Ulmus glabra ; but none of the figures here given enter sufficiently into such details, nor do they show that the leaves are doubly serrated. Although long naturalised in Britain the elm comes from the East. Loudon quotes an opinion that it was brought back by the Crusaders. The species are all much alike, and were considered by Linnæus to be mere varieties of one. They rise to from 40 to 80 feet in height, but no greater contrast could there be than between the Common Elm and the Wych Elm in general appearance.

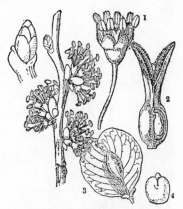

781.—FLOWERS AND FRUIT OF ULMUS CAMPESTRIS. L. V. K. 580.
Purplish. Mar.—April.

782.—CORK-BARKED ELM.
ULMUS SUBEROSA (above).
WYCH ELM.
ULMUS MONTANA (below). Z. 200.

783.—COMMON ELM.
ULMUS CAMPESTRIS.
M. (1565) 144.

784.—DUTCH ELM.
ULMUS MAJOR.　D. 89.2.

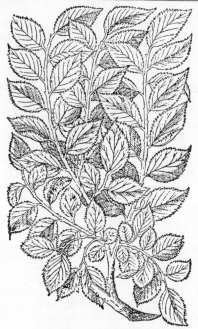

785.—WYCH ELM.
ULMUS MONTANA.　G. 1297.2.

CANNABINACEÆ—HOP AND HEMP.

"Herbaceous rough-stemmed watery plants with alternate lobed stipulate leaves, and small inconspicuous flowers."—Lindley. The plants are diœcious, the male flowers in panicles, the female in a form resembling a cone. In the annexed figure, from Lindley, (1) is a male flower and (2) a female flower, which has a single sepal like a cloak enclosing the ovary. The similarities of this cone-form (strobile is the botanical word) to the catkin cannot pass unobserved—we are, indeed, approaching the plants with catkins. In Hop the leaves are opposite, and very vine-like. HEMP is a plant of Indian origin, cultivated for its fibre, of which canvas and rope are made, and for its seeds, which produce oil, and are used as bird-food.

786.—HOP.
HUMULUS LUPULUS.　L. V. K. 265.

787.—HEMP.
CANNABIS SATIVUS. F. 393.
Green. 6 ft. June—July.

788.—HOP.
The Female Plant. C. E. 933.
Catkins, or strobiles, 1 in. July.

789.—HOP.
The Male Plant. Z. 297.2.
Leaves 4 inches across.

790.—HOP.

Humulus Lupulus.　F. 164.

The imbricated flower-heads, or strobiles, are the "hops" which are used to impart to beer its tonic and soporific qualities. It is interesting to compare this drawing with those of Gesner, Figs. 788 and 789, and with that of Lindley, Fig. 786. Note particularly the separated calyx at the top right-hand side of Fig. 788, and compare it with No. 2 in Fig. 786.

URTICACEÆ.

An Order akin to Cannabineæ and producing plants yielding fibre, and well known to us through the familiar stinging nettles. Of the three common kinds, the Small Nettle, Urtica urens, is about 1 foot high; Roman Nettle is about 2 feet, and the Common Nettle is 2 to 4 feet high. All are stinging. The flowers are well shown in Fig. 792; the male flowers cross-like with 4 stamens, the female in globular heads. All flower about July.

791.—COMMON NETTLE.   792.—ROMAN NETTLE.

URTICA DIOICA. F. 107.   URTICA PILULIFERA. F. 106.

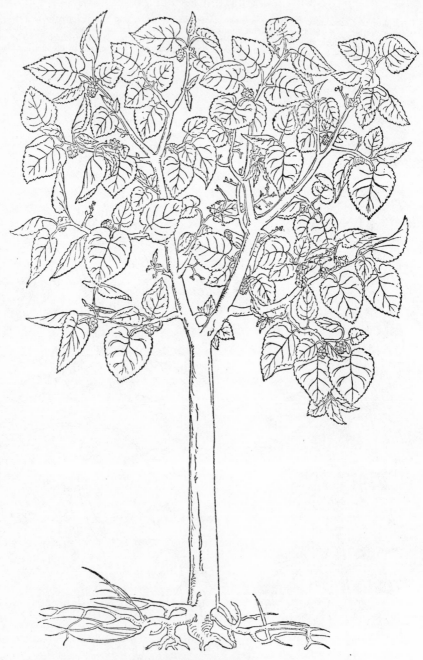

793.—MULBERRY [Moraceæ].

Morus nigra.   F. 522.

Fruit white or pink.   30 ft.   Flowers in June.

The details here both as regard leaf and berry are too large.   Introduced into England from Italy in 1548.

MORACEÆ.

Trees and shrubs with a milky juice, which in some cases yields India-rubber, as do the juices of certain plants of the Euphorbiaceæ. The flowers are unisexual and not unlike those of Nettles. Fruit, small nuts enclosed within a succulent receptacle, as the Fig, or within the consolidated succulent calyxes, as in Mulberry. The "grains" of the Mulberry, at first sight like those of Blackberry, are seen on examination to be composed of 4 tightly-closed fleshy sepals, or parts of the calyx, 2 within 2.

The Sycamore, or Pharaoh's Fig, is grown in Egypt for its fruit, and for the shade it casts.

794.—MULBERRY.
Morus nigra. L. ii. 196.1.
Pink or white. Fruit 1 in.

795.—FIG.
Ficus Carica. L. ii. 197.2.
15 ft. June—July.

796.—SYCAMORE FIG.
Ficus Sycamorus. L. ii. 197.1.
40 ft.

PLATANACEÆ—PLANE.

The leaves are alternate, with deciduous sheathing
stipules. The flowers in globular catkins hanging down, 2
or 3 on a stalk. These trees, which are from Barbary, the
Levant and North America, are amongst the handsomest of
our trees, and greatly beautify the squares of London. The
bark is deciduous.

797.—PLANE-TREE.
PLATANUS.   C.E. 62.

EUPHORBIACEÆ—SPURGES.

Herbs, shrubs, and trees, generally with an acrid, milky
juice, usually, and sometimes very, poisonous.

This is an Order remarkable for the instability of its
characteristics. It is generally considered as apetalous, but
some plants have evident petals. To the draughtsman there
is also little similarity between Box, Castor-Oil Plant, the
Spurges of the fields, and Mercury.

Perhaps the most important and constant peculiarity is
the fruit, which commonly is tricoccous, consisting of 3
carpels, which split and separate from their common axis.
But sometimes it is of 1, 2, or many cells, instead of 3.
It is generally globular, and is crowned with the styles,
which as a rule indicate by their number the number of the
cells. But if the plants vary greatly, they sometimes are
unmistakable. Particularly so is Sun Spurge, and its like.

798.—"MYRTLE SPURGE."
"TITHYMALUS MYRTIFOLIA."   G. 402.3.

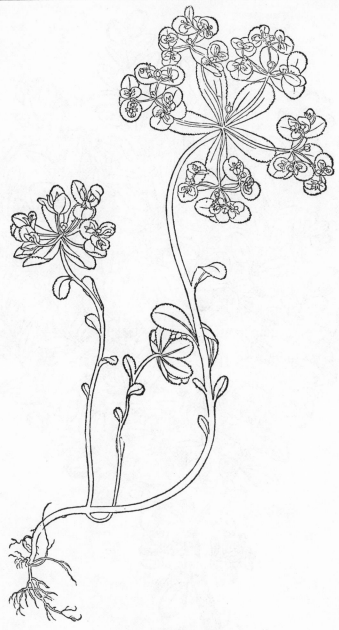

799.—SUN SPURGE.
Euphorbia Helioscopia. F. 811
1 ft. A common, yellowish-green plant.

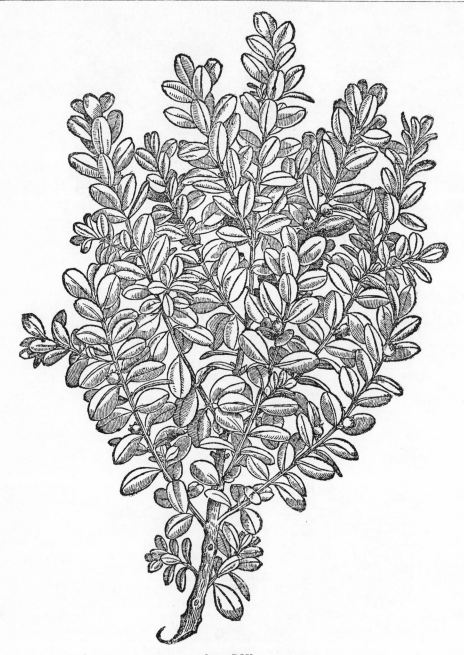

800.—BOX.

Buxus sempervirens. M. (1565) 190.

An evergreen shrub. Leaves convex, glossy. Wild upon chalk hills in certain localities in England. 3 to 15 ft. Flowers yellowish. April. The plant is here shown in fruit.

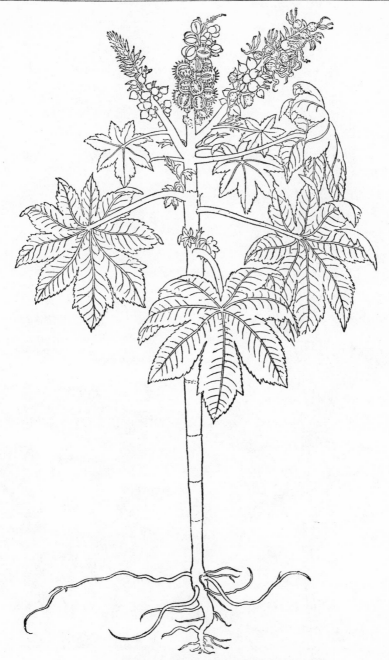

801.—PALMA CHRISTI, or CASTOR-OIL PLANT.
RICINUS COMMUNIS. F. 340.

In our gardens this is an annual rising to 6 ft. high, but in Southern Italy and Africa it becomes a moderate-sized tree. The female flowers are seen at the top with their red styles like the rays of star-fish. Below are some unopened buds, and below them apparently are some male flowers, the very numerous stamens represented as rough balls. On the middle stalk are the 3-capsular fruit with their prickles. A good figure is given in Curtis's *Botanical Magazine*, No. 2,209, where this present figure is commended. Flowers in July and August. The plant is held to be the "gourd" of Jonah.

405

803.—ENGLISH MERCURY.
MERCURIALIS PERENNIS. F. 444.
The female plant. Green. 1–2 ft. April—May.

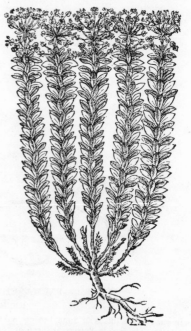

802.—FRENCH MERCURY.
MERCURIALIS ANNUA. F. 473.
The male plant. The fertile flowers are axillary, 2 together.
1 ft. Flowers green. Aug.

804.—"PHISICK SPURGE."
"TITHYMALUS MYRSINITIS." G. 402.4.

EMPETRACEÆ—CROWBERRY.

Small heath-like evergreen shrubs.

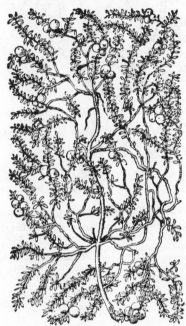

805.—CROWBERRY.
EMPETRUM NIGRUM. D. 188.1.
Purplish. Berries black. 6 in. April—June.

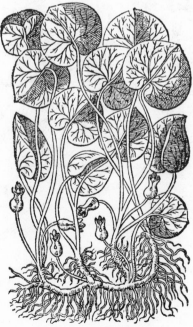

806.—ASARABACCA.
ASARUM EUROPÆUM. G. 688.1.
Green-brown. 4 in. May.

ARISTOLOCHIACEÆ—BIRTHWORT.

An Order of herbs or climbing plants having flowers with a tubular calyx superior to the ovary. The stamens and style are hidden within the flower.

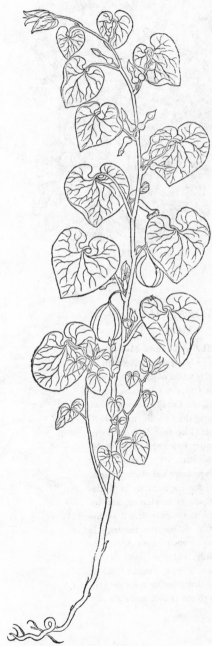

807.—SLENDER BIRTHWORT.
ARISTOLOCHIA CLEMATITIS. F. 90.
Pale yellow. 2-4 ft. Fl., ¾ in. July—Sep.

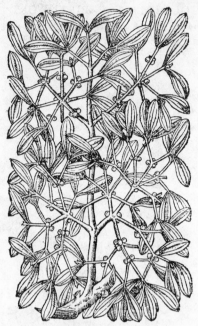

808.—MISTLETOE. G. 1168.1.

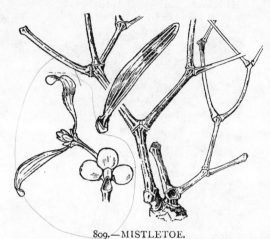

809.—MISTLETOE.

Says Gerarde—"Birdlime" [the glue made from the "Missell"] "is hot and biting, and consisteth of an airie and waterie substance, with some earthie qualitie." "This Birdlime inwardly taken is mortall, and bringeth most greevous accidents, the toong is inflamed and swolne, the minde is distraughted, the strength of the hart and wits faile."

LORANTHACEÆ— MISTLETOE.

The well-known Mistletoe is the only representative with us of this Order of parasitic shrubs. The Order has been associated with the Caprifoliaceæ and put among the Corollifloræ—it is so placed in the *British Flora*. Bentley followed Lindley in denominating its flowers Monochlamydeous and placing it here. The flowers, certainly in Mistletoe, are inconspicuous, and indeed no designer dare draw the plant without the berries. It grows upon the Apple-tree chiefly, but also upon Maple, Lime, Willow, Fir and Oak. It was that growing upon the Oak which the Druids made use of.

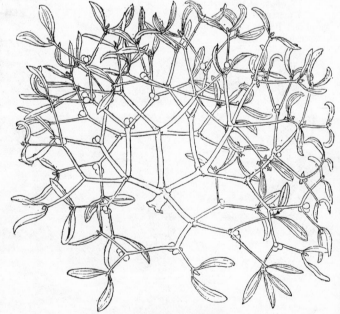

810.—MISTLETOE.
VISCUM ALBUM. F. 329.
Berries ½ in., white.   Leaves 2 in.

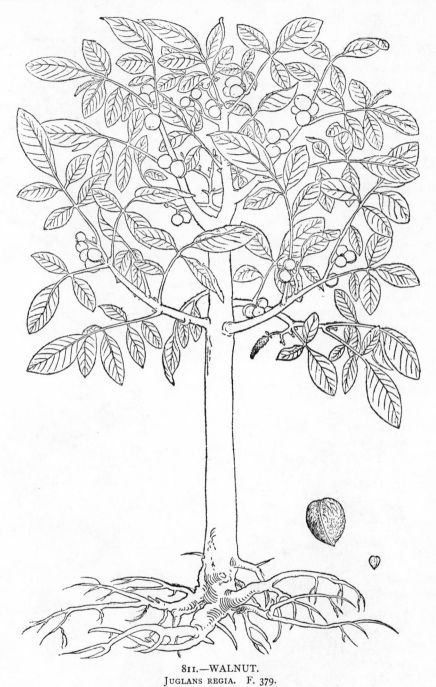

811.—WALNUT.
JUGLANS REGIA. F. 379.

The leaves are alternate, pinnate, without stipules. The male flowers are in catkins, the female flowers two or three together. Native in Persia, Cashmere, etc. 50 ft. April—May.

CORYLACEÆ—HAZEL—OAK—BEECH

The Order is also called Cupuliferæ. The cupule is the husk in Hazel, or the cup in Oak. It is the involucre surrounding or enclosing the female flowers. The male flowers are clustered together into an *amentum* or catkin, which consists of scales closely arranged upon a central common stalk, into a cylindrical form, and each enclosing one or more flowers. Catkin means a *little cat*, and alludes to the similarity of the inflorescence to a cat's tail.

The leaves are alternate, with deciduous stipules, and generally have their veins fairly parallel, running right out from the mid-rib to the border.

Of COMMON BRITISH OAK, Quercus Robur, there are two varieties, one of which has long fruit-stalks or peduncles, and therefore called Q. pedunculata. It has sessile leaves. The other has its flowers without peduncles, and is therefore called Q. sessiliflora. Its leaves (to preserve contrariness) have long stalks, and are wider than the leaves of the other kind.

812.—QUERCUS ROBUR SESSILI-
FLORA. D. 2.2.

813.—OAK WITH MOSS.
QUERCUS SESSILIFLORA. L. ii. 155.1.

814.—OAK WITH GALL-APPLE.
QUERCUS PEDUNCULATA. L. ii. 154.2.

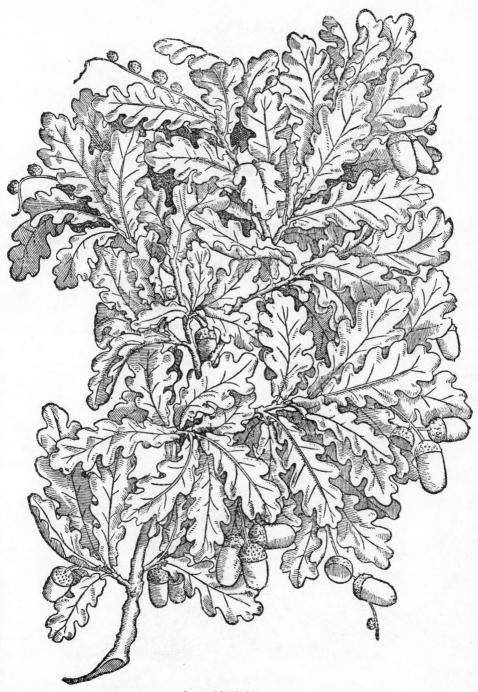

815.—COMMON OAK.

Quercus Robur pedunculata.　M. (1565) 204.

816.—COMMON OAK.

Quercus Robur sessiliflora.　F. 229.

817.—KERMES OAK.

Quercus coccifera.　M. (1565) 1030.

Height, 10 ft. From this are collected the kermes or scarlet grains, a little red gall occasioned by the puncture of the Coccus ilicis. From these kermes scarlet dye was produced till they were superseded by similar galls produced on the Cactus by the Cochineal insect. All have now given place to aniline dyes. The Kermes Oak is native in Southern France.

818.—KERMES OAK.
QUERCUS COCCIFERA. J. 1342.

819.—KERMES OAK.
QUERCUS COCCIFERA. D. 27.

820.—KERMES OAK.
QUERCUS COCCIFERA. M. 460.

821.—ITALIAN OAK.
QUERCUS ESCULUS. D. 5.1.
Acorns sweet, eaten in S. Europe.

822.—TURKEY OAK.
QUERCUS CERRIS. J. 1348.2.
Height, 50 ft.

823.—TURKEY OAK.
QUERCUS CERRIS. J. 1345.2.

824.—HOLM OR HOLLY OAK.
QUERCUS ILEX LATIFOLIA. C. E. 114.
Height, 60 ft.

825.—EVERGREEN OAK.
QUERCUS ILEX. J. 1344.1.

826.—EVERGREEN, HOLLY OR HOLM OAK.
QUERCUS ILEX. M. (1565) 206.

827.—CORK OAK.
QUERCUS SUBER.　M. (1565) 207.
Bark cracked, fungous.　Height, 20 ft.　South Europe.

828.—EVERGREEN OAK IN FLOWER.
Quercus Ilex. J. 1344.2.

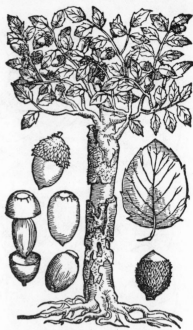

829.—CORK OAK.
Quercus Suber. Z. 161.1.

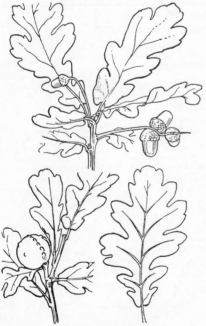

830.—OAK and OAK GALL.
Quercus pedunculata.

831.—HORNBEAM.
Carpinus Betulus. Z. 201.1.
Brownish-green. A tree. April—May.

832.—HAZEL.
CORYLUS AVELLANA.　M. (1565) 281.
Catkins yellowish, stigmas crimson.　A large shrub.　The nuts with the long cupules are
Filberts, those with the short cupules are also called Cob-nuts.

833.—BEECH.
FAGUS SYLVATICA. M. (1565) 205.
Catkins brownish. 70 ft. April—May.

834.—SWEET OR SPANISH CHESTNUT.
CASTANEA VULGARIS, or VESCA. M. (1565) 211.
Fertile flowers three together, and barren flowers on long pendulous stalks ; yellow, fragrant. See p. 528.
Height, 50 ft. May—June.

835.—SWEET OR SPANISH CHESTNUT.
CASTANEA VULGARIS, or VESCA. F. 377.

The cupules are here shown, large and spiny. The long male catkins, sweet-smelling, are very conspicuous—
see last figure. In neither case are the female flowers shown, but only when in seed.

MYRICACEÆ—GALE.

An Order of Amentaceous shrubs with alternate leaves.
SWEET GALE is the only British species. The leaves are
lanceolate and covered with white resinous glands. Scales
of the catkins pointed.

BETULACEÆ—BIRCH—ALDER.

These plants are similar to the Corylaceæ, but both
male and female flowers are in catkins—the male always
the larger. The leaves are alternate, with deciduous stipules, and have the veins running right out to the margin.

The leaf of the Alder inclines to that of the Hazel, but
has an abrupt point, suggesting a slight notch. The leaf of
the Birch is of the stiffer character seen in Chestnut and
Oak. There are no cupules or cups to the fruit. The fruit
of Alder is a cone.

836.—SWEET GALE.
MYRICA GALE. J. 1414.
Yellowish-red. 4 ft. May.

837.—ALDER.
ALNUS GLUTINOSA. C. E. 68.
Male catkins reddish. March—April.

838.—ALDER.
ALNUS GLUTINOSA. M. (1565) 140.

839.—BIRCH.
BETULA ALBA.　M. (1565) 142.
Bark silvery white.　40 ft.　April—May.

SALICACEÆ—WILLOW, POPLAR.

These are diœcious plants, as the Betulaceæ are monœcious. The catkins in Willow are erect, and from he hairiness or furriness of their scales are by children called "pussy-willows." The branches bearing them also serve as "palms" on Palm Sunday. As the catkins advance, the male are seen to put out their stamens with yellow anthers, and the female to become covered with green bottle-shaped fruit. In Poplar the scales are also furry, but the catkins are limp and hang down.

The leaves are of graceful outline—both of Willows and Poplars, and have stipules. Their veins melt away, and do not run out to the edges as they do in the other Amentaceæ. The edges are finely serrated.

The flattened leaf-stalks in Poplar must not be overlooked. Their being flattened facilitates their trembling. A large-leafed Poplar seen in gardens is the Balsam Poplar.

In all Amentaceous plants the catkins appear before the leaves, and are wind-fertilized.

Of Willows no fewer than 76 species have been enumerated as British, though in the *British Flora* the number given is 15.

(See Appendix, Fig. 53.)

840.—CRACK WILLOW.
SALIX FRAGILIS. D. 276.3.
Showing the stipules.

841.—GOAT WILLOW.
SALIX CAPRÆA. J. 1390.2.

842.—YELLOW WILLOW.

SALIX ALBA. F. 335.

Branches yellow. . It is a variety of Salix alba, and sometimes called Salix vitellina—the Golden Osier.
10 to 30 ft. high.

843.—ASPEN.
Populus tremula.　M. (1565) 138.
Height, 50 ft.

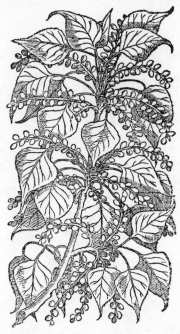

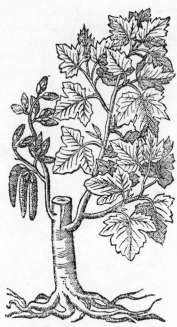

844.—POPLAR.
Populus nigra. M. (1554) 88.2.
30 ft.

845.—ABELE.
Populus alba. L. ii. 193.1.
40 ft.

[Here end the *Monochlamydeous* Exogens, and here begin the *Gymnospermous* Exogens—those which have their ovules naked upon carpellary scales, and not enclosed in carpels.]

CONIFERÆ—PINE TREES.

Trees or evergreen shrubs, with an upright branched trunk, abounding in resin. The leaves are linear, th entire margins, sometimes in bundles. The female flowers are in cones, which consist of overlapping scales, which are the ovaries or carpels. They are without style or stigma, and bear two or more naked ovules. The scales are in the axils of bracts, which are particularly evident in the young rose-purple cones of the Larch, where they are narrow, green, and pointed. The cones become enlarged and hardened. The male flowers "consist of a single stamen, or of a few united, collected in a deciduous amentum."

See a figure of the cone, or "pine apple," in the Appendix, Fig. 57.

The Scotch Fir is hardly well represented in Fig. 855. The Scotch Fir is remarkable for its red naked stem and its contorted arms, making a strong contrast to the formality of the common fir. The lower branches frequently decay and fall off, and in old trees the mid-branches hang gracefully pendent. The central stem ultimately turns over, as if bent by the wind, and the head of the tree, where the small branches are free and plentiful, makes a dense mass of dark green. The leaves are denser and shorter than in the other firs. The cones are small and pointed.

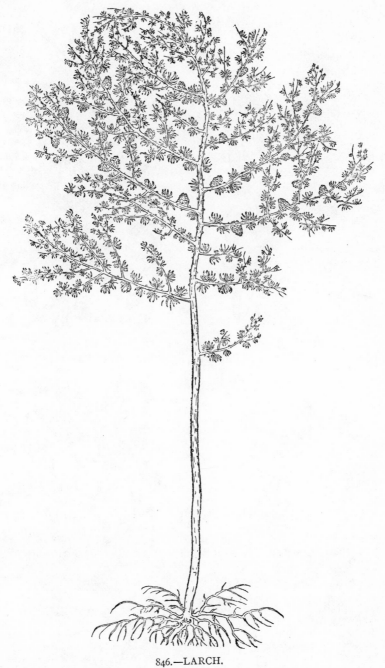

846.—LARCH.

PINUS LARIX. F. 496.

    The leaves in little tufts.   The tree assumes a conical head ; here the head is curved down, probably to get it in the block.

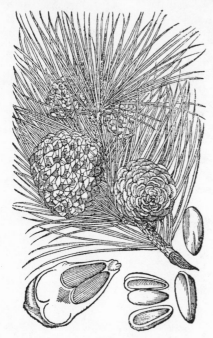

847.—STONE PINE.
PINUS PINEA. C. E. 39.

848.—SPRUCE.
ABIES EXCELSA. Z. 182.1.

849.—SILVER FIR.
PICEA PECTINATA. Z. 182. 2.

850.—SILVER FIR.
PICEA PECTINATA. Z. 183.

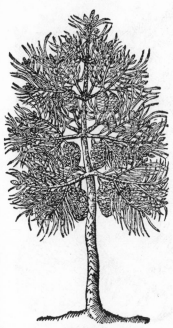

851.—STONE PINE.
Pinus Pinea. J. 1355.

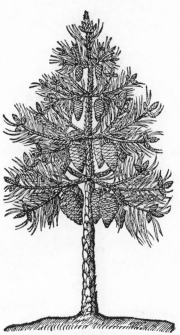

852.—"SMALLER WILD PINE."
L. ii. 229.

853.—FIR TREE.
Picea. J. 1363.1.

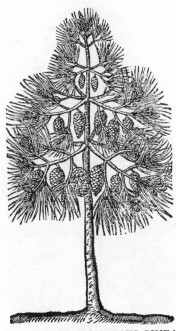

854.—"WILD MOUNTAIN PINE."
J. 1357.3.

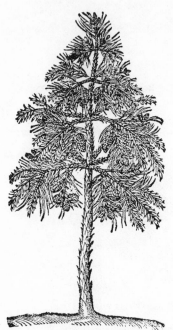

855.—SCOTCH FIR.
PINUS SYLVESTRIS. J. 1356.1.

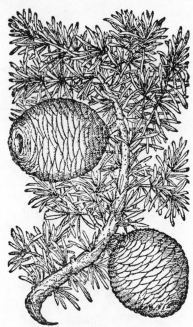

856.—CEDAR.
CEDRUS. D. 36.2.

857.—CYPRESS.
CYPRESSUS SEMPERVIRENS. Z. 106.

858.—CYPRESS.
CYPRESSUS SEMPERVIRENS. D. 58.

TAXACEÆ—YEW.

An Order differing from the Coniferæ in the construction and form of its flowers and fruit. The globular forms in the figure suggest, in being round, the yellowish male flowers which occur on the backs of the leaves. The pink bells of the fruit make a pretty effect upon the plant when they are plentiful.

The Upright or Irish Yew, Taxus fastigiata, rises in a pointed or fastigiate form after the manner of the Lombardy Poplar. Its leaves are not disposed featherwise, but are scattered around the stems, though in one variety, Taxus erecta, the leaves are 2-ranked as in the Common Yew, but the branches are upright.

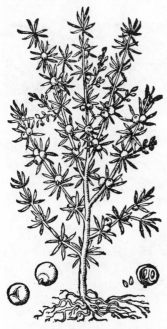 

859.—JUNIPER [CONIFERÆ].
JUNIPERUS COMMUNIS. Z. 107.1.
Catkins brownish. Berries green or black. May.

860.—YEW.
TAXUS BACCATA. J. 1370.
Fl., yellowish. Fruit a naked green seed within a succulent pink bell. March—April.

[Here end all the EXOGENS, and here follow the DICTYOGENS.]

DICTYOGENS.

Lindley placed between the Exogens and Endogens certain Orders which he classed under the above name. They are now mixed up by botanists with the Endogens or Monocotyledones ; but for our purposes this is the proper place for some, at least, of them.

DIOSCOREACEÆ.—For a brief general description, see p. 434.

SMILACEÆ.—For a brief general description, see p. 436.

TRILLIACEÆ.—Represented in Britain by Herb Paris. The leaves are whorled, net veined, but not articulated to the stem, and therefore nearly approach Endogens. Flowers with an inferior perianth of 6 or 8 parts, in 2 rows, the inner larger. For Herb Paris, see Appendix, Fig. 54.

DIOSCOREACEÆ—YAMS.

Diœcious. Twining shrubs with large tubers, or yams, above or below ground. Leaves alternate, with reticulated veins. Calyx and corolla confounded in a perianth of 6 points. Male flowers with 6 stamens. Female flowers with 3 styles, or 1, which is trifid.

861.—BLACK BRYONY.
TAMUS COMMUNIS.　M. (1565) 1285.
Green.　Berries scarlet.　Leaves, 5 in.　Climbs over hedges.　June.

862.—ROUGH BINDWEED [SMILACEÆ].

SMILAX ASPERA. F. 718.

The flowers here shown with 5 instead of 6 segments to the perianth. South of Europe.

SMILACEÆ—SARSAPARILLA.

Plants, herbaceous or scarcely woody in stem, and tending to climb ; chiefly from tropical America. The flowers have a perianth of 6 parts, 3 within 3, all alike, 6 stamens, a trifid style surmounting a 3-celled ovary. The leaves have strong longitudinal nerves, with reticulated venation between. Fruit baceate.

863.—ROUGH BINDWEED.
Smilax aspera. D. 1422.

864.—ROUGH BINDWEED.
Smilax aspera. G. 710.3.

[Here end the Dictyogens, and here follow the Endogens, and first the *Petaloid* Endogens, which have whorled, coloured, perianth.]

ORCHIDACEÆ—ORCHIDS.

Herbs or shrubs readily recognized by their peculiar flowers, which stand in the axils of small bracts upon short twisted stalks, which are the ovaries. In temperate regions Orchids are terrestrial herbs sending up a straight stalk which has a few leaves below, and bears above a number of flowers in a spike or raceme. In the Tropics the plants are shrubby and grow upon trees, or fix themselves upon stones. These exotic species are very largely cultivated in greenhouses. They had not been discovered at the time when the ancient herbals were composed, and consequently the Orchids figured in those books are of the much humbler kind which is native in Europe.

The bracts sometimes are twisted around the ovaries, as in Fig. 865, and appear to be part of the flower. The perianth is of 6 parts, commonly called 3 sepals and 3 petals. The 3 sepals are generally somewhat long and narrow—one is directed upwards and two horizontally sideways, like wings. Two of the petals are similar to them and are alternate with the sepals. The third is very different, usually lipped, and spurred. The lip is somewhat notched at the margin, sometimes very deeply, so deeply as to even suggest by its long segment the limbs of a man. Indeed in no flowers are there so many names of other creatures—the Man, the Military, the Bee, the Fly, the Frog, the Lizard, the Spider, etc.

The organs of the flower constitute a "column," consisting of 3 stamens, of which the middle one alone is perfect, consolidated with the style into a central body, so that the style is next to the lip and the stamens are next to the back sepal.

The forms of some of the flowers are given in the Appendix, Fig. 54.

The leaves are very simple, and sheath the stalk at the base.

Lady's Slipper, Cypripedium Calceolus, and Brassia (see Appendix, Fig. 54) belong to this Order.

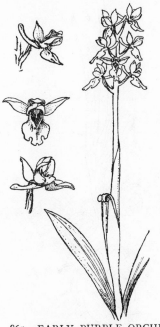

865.—EARLY PURPLE ORCHID.
ORCHIS MASCULA.
Purple. 1 ft. April—May.

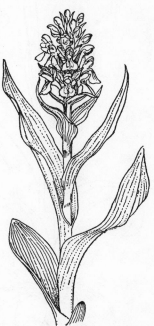

866.—MARSH ORCHID.
ORCHIS LATIFOLIA.
Pink. 1 ft. May—June.

867.—ORCHID.
F. 557.

868.—BEE ORCHID.
OPHRYS APIFERA. F. 560.
Calyx pink-purple, lip purplish brown with
yellow lines. 9 in. June.

869.—DWARF DARK-WINGED ORCHID.
ORCHIS USTULATA. L. i. 184.2.
Purplish brown, lip white and spotted.
5 in. June.

870.—SMALL BUTTERFLY ORCHID.
HABENARIA BIFOLIA. G. 162.1.
Greenish white. 10 in. June—July.

871.—BIRD'S-NEST ORCHID.
NEOTTIA NIDUS-AVIS. L. i. 195.1.
Brown. 10 in. Scales brown. May—June.

872.—LADY'S TRESSES.
SPIRANTHES AUTUMNALIS. L. i. 186.2.
White. 9 in. Aug.—Sep.

873.—BANANA. ADAM'S APPLE.

Musa paradisiaca? J. 1515.1.

MUSACEÆ—BANANA.

" Musa paradisiaca rises with a soft herbaceous stalk 15 or 20 feet high, with leaves more than 6 feet long and 2 feet broad. When the plant is full grown the spike of flowers appears from the centre of the leaves and nods on one side."—Loudon.

The fruit which succeeds is the well-known Banana, in the dense "crowns" now so familiar. The plant was held to have been the Forbidden Fruit of Paradise, and also the Grapes brought out of the Promised Land by Joshua and Caleb. Also called Plantain.

IRIDACEÆ—IRIS, CROCUS, Etc.

Perhaps the chief peculiarity of the Iris and Crocus is the petal-like stigmas. The style rises up (unseen in Iris), and then divides into 3 petal-like portions, which carry stigmatic surfaces. In Iris these 3 stigmas are arched in their longitudinal direction and roof-like in section. They terminate in 2 wings curled upwards. In Crocus the 3 stigmas remain upright. (See the Appendix, Fig. 54.)

The ovary is below the flower, from which it is separated by what may be called the tubular part of the perianth. In Crocus this " tube " is so long that the ovary is hidden low down in the spathes, amongst the leaves.

The bracts which enclose the flower are usually spathaceous and generally scarious—dry and membranous. Their colour thus resembles parchment, or mummy-cloth, and their wrinkled character contrasts with the swift fresh lines of the flower.

In the annexed diagram are given the details of the form. A, B are sections of the ovary, which shoulders in, D, to a narrower tube-like part G, seen in section at E. The ovary is usually 3-cornered, as A, B, where it will be noticed there are nine ridges. Sometimes the ovary is more truly cylindrical. In the middle of the section E is seen a little trefoil—the style, which spreads out into the 3 stigmas, M. Each angle E if cut away carries with it a drooping sepal, H, and one of the 3 stamens, I. It is over the top of this stamen that one of the stigmas fits like a roof. Between the angles E E is a portion supporting a petal, L, which is sometimes very small, sometimes very large.

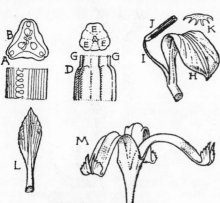

874.—DETAILS OF FLOWER OF IRIS.

The drooping sepal sometimes carries a " beard," as is seen in Fig. 879.

In colour Irises or Flags are yellow, pale blue, livid, white or purple. They are usually variegated, and the sepals are much gayer than the petals, which are not subject to variegation. The sepals are often marked with brown or purple in strokes following the direction of the " veins."

All the flowers are usually yellow in the depth, where the parts unite. The beard is yellow or pale. The sepals are darker than the petals—when they differ at all.

The 6 segments of the perianth of the Crocus are virtually 3 within 3—3 forming an outer covering or calyx. They are sometimes, like those of the Iris, striated with brown or purple.

The sword-like leaves of the Iris are cleft on their inner edge towards the base so that they allow the development of other leaves, and of the stem within them.

The leaves of the Crocus are remarkable for their growing to a considerable length after the flower has fallen.

Another plant sometimes with a regular perianth is the Morea. Gladiolus has also a six-parted perianth, but it is irregular and forms two lips. Its stem is 5 to 8 flowered. Each flower is accompanied by 2 bracts one of which has enclosed the other, and that other has enclosed the flower. Their edges curl inwards and make them awl-shaped. The lower segments of the perianth bear marks somewhat like the letter V, or else passing longitudinally down the middle. The flowers are red, rose, rose-purple, or white with marks of these colours. (See Appendix, Figs. 32 and 54—in the former the leaves are well shown.)

875.—DOUBLE-BEARING IRIS.
IRIS SUB-BIFLORA. L. i. 62.2.
Sepals purple, petals violet. 1 ft. Fl., 4 in.
April—May.

876.—PURPLE IRIS.
IRIS GERMANICA. L. i. 59.1.
Sepals purple, petals blue. 3 ft. Fl.,
5 in. May—June.

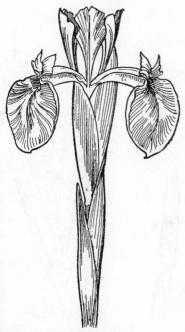

877.—GLADDON.
IRIS FŒTIDISSIMA.
Lead-colour. (See Figs. 885, 886 and 887.)

878.—DWARF IRIS.
IRIS PUMILA. C. E. 4.
Purple or yellow. 6 in. April—May.

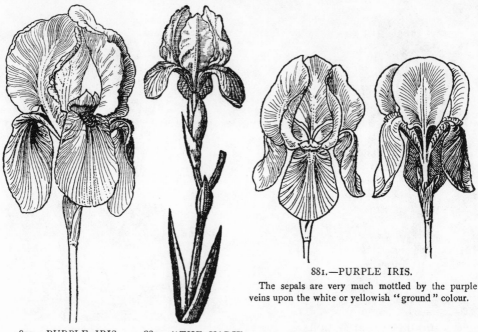

879.—PURPLE IRIS. 880.—"THE VARIE-
GATED YELLOW
IRIS."

881.—PURPLE IRIS.
The sepals are very much mottled by the purple
veins upon the white or yellowish "ground" colour.

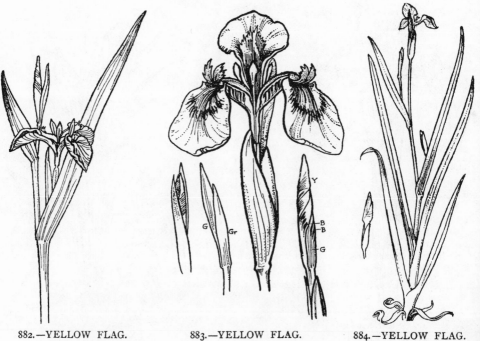

882.—YELLOW FLAG. 883.—YELLOW FLAG. 884.—YELLOW FLAG.
Iris Pseudacorus. The inner segments or petals very small. Yellow ; the sepals sometimes
marked with brown. July.

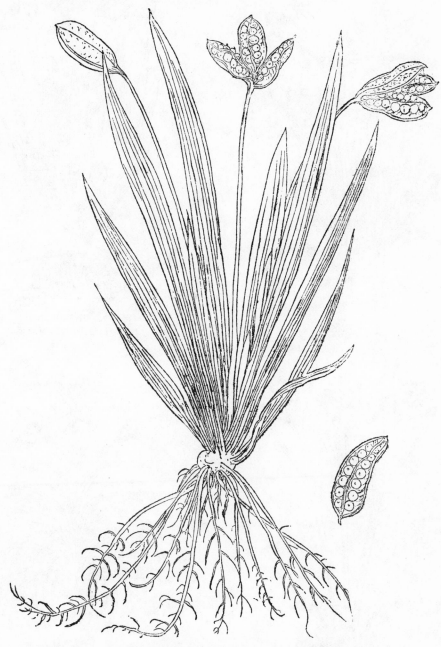

885.—STINKING GLADDON, or GLADWYN.
Iris fœtidissima. F. 794.

Flower (for which see Fig. 886) an "overworn bluish colour declining to grey or an ash-colour." Seeds orange-scarlet, remaining on the opened capsule for some time. Height, 2 ft. Shady places. June—July.

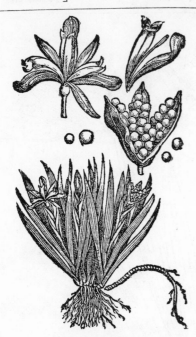

886.—GLADDON.
Iris fœtidissima.   Z. 390.
Livid or lead-colour.   2 ft.   June—Aug.

887.—GLADDON.
Iris fœtidissima.   M. 446. 1.
Berries orange-scarlet.

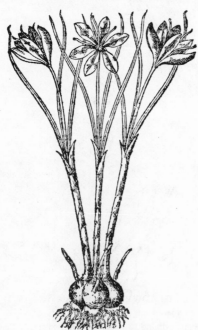

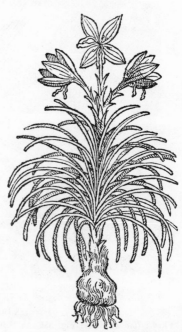

888.—SAFFRON CROCUS.
Crocus sativus.   G. 123.2.
Purple, stigmas orange.

889.—SAFFRON CROCUS.
Crocus sativus.   M. 47.
Stigmas hanging out.   9 in.   Sep.

AMARYLLIDACEÆ—DAFFODIL, Etc.

The perianth is of 6 segments, which are much alike, but the 3 outer are slightly greener and calyx-like than the inner, which are broader and lack all greenness. The bracts enclosing the flowers are generally spathaceous, dry, and of the colour of mummy-cloth. Many species have a *corona* in the form of a cup or tube surmounting the tube of the perianth ; these are species of Narcissus, and the common Daffy-down-dilly is the best known. Snowdrop (see App., Fig. 34), Snowflake (Fig. 892), the various kinds of Daffodil and Narcissus (Figs. 890, 891, and App., Figs. 35 to 39), Amaryllis (App., Fig. 53), and the American Aloe (Fig. 893), and Pancratium belong to this Order. The stamens are 6, the style 1, the stigma 3-lobed, the leaves sword-shape.

890.—"PRIMROSE PEERLESS."
G. 110.6.

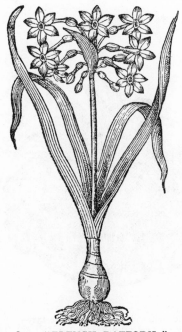

891.—"FRENCH DAFFODIL."
G. 110.7.

892.—SNOWFLAKE.
LEUCOIUM ÆSTIVUM. C. E. 957.
White, green mark. 9 in. May.

893.—AMERICAN ALOE [AMARYLLIDACEÆ].
AGAVE AMERICANA.   L. 374.1.
Flowers yellow in a tall upright raceme.
20 ft.   Aug.—Oct.

894.—PINE-APPLE.
ANANASSA SATIVA.   L. i. 375.1.

These are the tufts of leaves which occur at the top of the fruit, and are given because they are fine examples of strong, boldly-modelled form.   The PINE-APPLE belongs to the Order BROMELIACEÆ.   This fruit consists of the consolidated fruit and the bracts within which they have grown.   The uppermost bracts are sterile.   See Appendix, Fig. 53.

446

895.—TURK'S CAP LILY.
LILIUM MARTAGON. F. 115.
(See Appendix. Fig. 50.)
Colour varies from purple to whitish-purple
and white ; the colour deeper and perhaps
redder towards the base outside, spotted
with purple within. 3-4 ft.

LILIACEÆ—LILIES.

An Order of considerable variety and uniting plants of very
different appearance. Among them is, if we mistake not, the
only Endogenous shrub in our temperate regions—the Butcher's
Broom. So diverse are the plants of this Order that the marks
of a natural affinity seem obliterated. The main distinction
between them and the Amaryllidaceæ and Iridaceæ is in the
ovary being superior. This is important from the draughtsman's point of view, as he has to remember that there is no
green base to these flowers, but that the stalks hold the
perianth directly. The perianth is of 6 segments, either separate
or joined together into a bell or tube. In the Lilies and Tulips
they fall off. The stamens are 6, inserted upon the segments
of the perianth. Ovary 3-celled, style 1, stigma simple or
3-lobed.

The leaves are simple, usually gaining some breadth and then
drawing to a sharp point. They are generally strongly ribbed,
and they are not articulated to the stem, and do not fall off.

Of the 6 segments of the perianth, 3 are at the outside, preserve some greenness, and are rather narrower than the inner
3. In Tulip the margins of these sepals are folded in, in the
bud, which is green. When the flower opens this greenness
gradually disappears and red or yellow or white takes its place.

Several figures of plants of this Order are given in the
Appendix, Figs. 40 to 50.

A glance through the illustrations shows that the plants can
be ranked, as far as appearance is concerned, in several kinds.
Branching we only see in Butcher's Broom and Asparagus,
unless we go further afield to the tropical genera—where is
such a huge tree as the Dragon Tree. We see that bulbous
plants—and most of our petaloid Endogens, whether of this or
preceding Orders, are bulbous—in their normal condition send
up a single flowering-stalk, with, at its base, or upon its lower
part a number of leaves of the very simplest form. Each
flowering-stalk is in this way surrounded and accompanied by
its leaves. And when there are many flowering stalks rising
from one root they still have their own separate leaves as if they
were merely a number of little plants closely set side by side.
This flowering-stalk does not branch. At the top it bears one
flower, as in Tulip and Crocus, but in other cases there are
several or many flowers arranged upon the upper part of the
stalk in several different ways. In the Lilies the flowers are in
the axils of the uppermost leaves, which are small enough there
to be called bracts. Sometimes 2 or 3 flowers are together at
the top. This arrangement takes very positive form in the
Alstrœmerias, which belong to the Amaryllidaceæ. In some
species of it there are several, say 6, little flower-stalks arranged
in a whorl at the summit, each bearing 2 flowers. A similar
arrangement is seen in the Martagon, Fig. 903.

It will not escape observation that the pedicels or short
flower-stalks in the Lilies are not always convex upwards, but
that they commonly have some measure of the contrary direction,
as is seen in a rather exaggerated manner in Fig. 905.

447

The Fritillaries have their flowers arranged much in the same way as Tulips and Lilies.

Hyacinth, Star of Bethlehem (Ornithogalum), Scilla, Lily of the Valley, and Anthericum have their flowers in racemes, which are sometimes one-sided, sometimes branched. The difference between Scilla and Ornithogalum may be almost said to be that the former are blue, and the latter white.

Other plants have their flowers in umbels, as Gagea, Garlic, but in the Onions the flowers are in round heads. The genus Allium, to which the Onion belongs, presents certain individual peculiarities. In Garlic, A. sativum, the umbel has bulbs at the bases of the pedicels and the leaves are broad. In Onion, A. Cepa, the leaves are tubular and the stem inflated, below the middle. In Chives the leaves are cylindrical. In Allium the flowers are enclosed in one or two spathes.

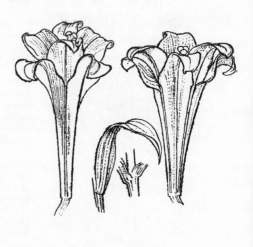

896.—"LILIUM CANDIDUM BYZANTINUM."
G. 146.2.
The lower flowers have the perianths fallen, exposing the ovaries and styles.

897.—LONG-FLOWERED LILY.
LILIUM LONGIFLORUM.
The segments are somewhat adherent by their edges.

898.—WHITE LILY.
LILIUM CANDIDUM. F. 364.
3 ft. June—July.
(See Fig. 900.)

899.—ORANGE LILY.
LILIUM BULBIFERUM. F. 365.
Orange. 4 ft. June—July.
(See Appendix, Fig. 48.)

900.—WHITE LILY.　MADONNA LILY.
LILIUM CANDIDUM.
Anthers yellow, style green, leaves wavy.　Height, 3 ft.　June—July.

901.—ORANGE LILY.
LILIUM BULBIFERUM.
L. i. 164.1.

902.—TURK'S CAP LILY.
LILIUM MARTAGON.
L. i. 168.1.
(See Fig. 895.)

903.—SCARLET MARTAGON LILY.
LILIUM CHALCEDONICUM.
L. i. 169.1.
Red. 4 ft. July—Aug.

904.—WHITE LILY.
LILIUM CANDIDUM. G. 146.1.

905.—"FIERY RED LILY."
LILIUM BULBIFERUM. G. 149.3.

(The plants do not branch as Fig. 904 is shown to do. It is one of the figures used by Sir E. Burne-Jones in the cover of *Studies in Both Arts*.)

906.—DAY LILY.
HEMEROCALLIS FULVA. G. 90.2.
Copper-coloured. 4 ft. Fl., 4 in. June—Aug.

907.—DAY LILY.
HEMEROCALLIS FULVA. J. 98.2.
The petals with wavy edges.

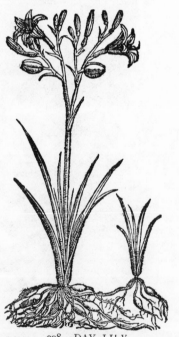

908.—DAY LILY.
HEMEROCALLIS FLAVA. J. 98.1.
Yellow. 2 ft. Fl., 3 in. June.

909.—"LAURUS ALEXANDRINA."
M. 524.1.

452

910.—BUTCHER'S BROOM.
RUSCUS ACULEATUS. M. (1565) 1214.

1–2 ft.   An evergreen shrub.   Leaves ovate, spine-pointed, bearing the flower on the surface.
Flowers greenish.   Berries red.   Common in woods.   March—April.

911.—ASPARAGUS. SPERAGE.
ASPARAGUS OFFICINALIS. F. 58.

The plant is properly erect and does not droop as shown in this beautiful drawing. Berries red. Flowers
yellowish-green, small, none here shown. July.

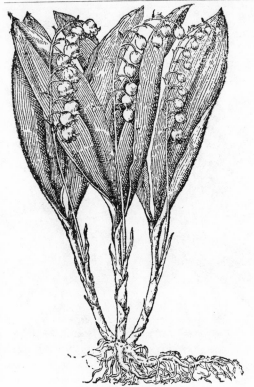

912.—LILY OF THE VALLEY.
CONVALLARIA MAJALIS. G. 331.2.
White, berries red.   8 in.   May.

913.—SOLOMON'S SEAL.
POLYGONATUM MULTIFLORUM. G. 756.3.
White, tipped green.   1–2 ft.   June.

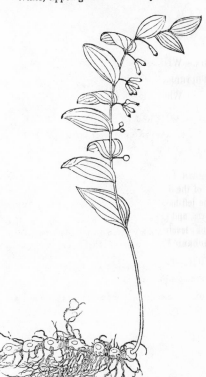

SOLOMON'S SEAL.—Few plants excel this in beauty.
The foliage is remarkably beautiful in line.   Fine as are
these drawings from Fuchsius, they do not quite do
justice to the plant.   The conventional (certainly as
regards its regimental stiffness) figure, Fig. 913, from
Gerarde has some measure of the flowing twist of the
leaves, which is hardly made enough of in the cuts from
Fuchsius.

LILY OF THE VALLEY.—The reader will have observed that the plants hereabout have bell-like flowers
and globular berries.

914.—COMMON SOLOMON'S SEAL.
POLYGONATUM MULTIFLORUM. F. 585.
White, tipped green.   1–2 ft.   June.

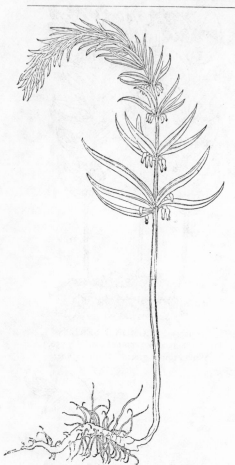

916.—PERSIAN LILY.
FRITILLARIA PERSICA.
G. 152.
Purple, tinged green. 2 ft.
Fl., 1 in. April—May.

917.—CROWN IMPERIAL.
FRITILLARIA IMPERIALIS.
G. 153.12.
Yellow. 4 ft. Fl., 2 in.
March—April.

915.—WHORLED SOLOMON'S SEAL.
POLYGONATUM VERTICILLATUM. F. 586.
White, berries blue. 2 ft. June.

CROWN IMPERIAL.—The flowers are just like those of the Fritillary given in the Appendix, Fig. 44, the left-hand example. They grow in the axils of leaves, and so close together as to be practically on one level. See the figure specified for the CHEQUERED FRITILLARY.

TULIP.—See the Appendix, Figs. 45, 46 and 47.

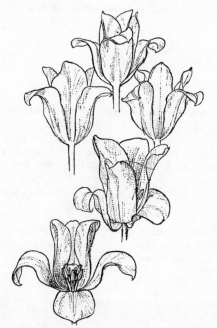

918.—TULIP FLOWERS.

919.—"RED AND YELLOW FOOLES COAT."
J. 144.24.

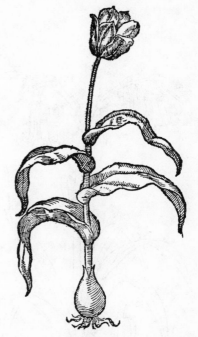

920.—"SULPHUR-COLOURED TULIP."
J. 144.25.

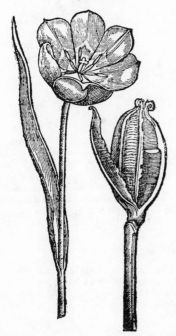

921.—"THE BLOOD-RED TULIP WITH
A YELLOW BOTTOM."
J. 139.8.

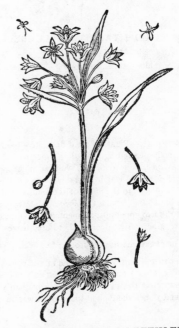

922.—YELLOW STAR OF BETHLEHEM.
GAGEA LUTEA.　D. 1502.3.
Fl., ¾ in.　March—April.

(See the Appendix, Figs. 40, etc., for several other plants of this Order.)

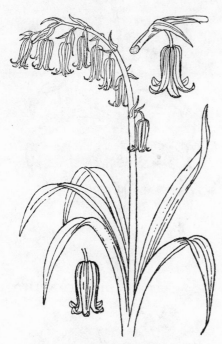

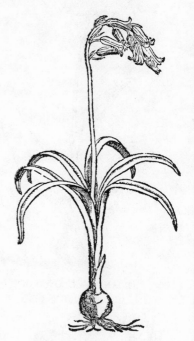

923.—WILD HYACINTH. BLUEBELL.
SCILLA NUTANS.
Blue or pink.   10 in.   Fl., ¾ in.   May—June.

924.—WILD HYACINTH.
SCILLA NUTANS.   L. i. 103.1.

ASPHODEL.—Plants of great decorative beauty, of which good figures are given in the Appendix, Figs. 1 and 49. Good coloured figures will be found in the *Botanical Magazine*, Nos. 773, 799 and 984. The "CANDY SPIDERWORT," Fig. 926, may well be an Asphodel. The greatest beauty in the Asphodels is the line of green, crimson, or purple which runs down the middle of each segment of the perianth, both inside and outside. It particularly adds to the effect of the buds. The flowers are yellow with a green line, or white with a crimson or purple line. Fig. 773 in the *Botanical Magazine* differs from the plant on the right of Fig. 49 in the Appendix, in that the five uppermost segments are all directed upwards, leaving one directed downwards, and in there being slender, rather curly, 3-cornered, striated leaves up the stems.

In height Asphodels are from 2 to 3 feet. They flower in May and June.

No better lesson could there be in the conventional rendering of plants than comparing the two figures in the Appendix, Figs. 1 and 49. It is certainly not easy to assert that the coarse woodcut used by Egenolph, about 1545, expresses the plant less effectively than the careful copper-plate of Crispin de Passe, done seventy years later, and done when plant-knowledge was in a very different condition.

925.—BRANCHED SPIDERWORT.
ANTHERICUM RAMOSUM. J. 48.1.
White. 2 ft. May—June.

926.—CANDY SPIDERWORT.
"PHALANGIUM CRETÆ." D. 1589.1.

927.—"TRUE SPIDERWORT OF THE
ANCIENTS."
J. 48.4.

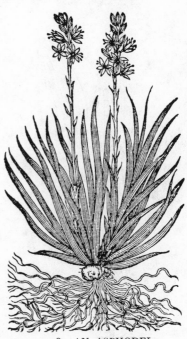

928.—AN ASPHODEL.
G. 88.2.

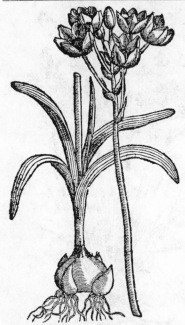

929.—"THE GREAT ARABICK STAR-
FLOWER. STAR OF BETHLEHEM."
ORNITHOGALUM UMBELLATUM. J. 167.7.
White. 1 ft. April—May.

930.—RAMSON.
ALLIUM URSINUM. J. 179.2.
White. 1 ft. April—May.

Very closely connected are STAR OF BETHLEHEM, Ornithogalum, STAR HYACINTH, Scilla, and the ordinary HYACINTH,
Hyacinthus. For figures of several of them see the Appendix,
Figs. 40, 42 and 43. Ornithogalum is white, Scilla is blue,
occasionally pink. The GRAPE HYACINTH is Muscari race-
mosum. The Wild Hyacinth or Bluebell (Harebell in Scotland) is a drooping Scilla. Another Scilla with 4 or 5 pendulous
flowers is common in gardens. One of the most beautiful of
the Stars of Bethlehem is Ornithogalum nutans, with drooping
flowers, which are green outside with white edges.

GARLIC. ONION. CHIVES.—These are species of Allium
and have already been mentioned on p. 448. The decorative
possibilities in the heads of flowers are great, and one regrets
that the figures in the herbals do not do justice to these plants.
The MOLY of Homer is the plant he says Hermes gave to
Ulysses to counteract the enchantments of Circe. The figure
here given, Fig. 931, compared with that in the *Botanical
Magazine*, No. 1148, has too spreading and sparsely flowered
a head. In the *Magazine* the centre, the ovary, is coloured
dark or black. The plant is remarkable for having a white
flower and a black root. A figure is also given in the Appendix,
Fig. 41. There also is given a Garlic with its bulbs clustered
at the bases of its pedicels. Its two bracts or spathes, which
are long and with slender points, are not, however, shown.

931.—HOMER'S MOLY?
ALLIUM MAGICUM.　G. 144.3.
White, greenish or purplish behind.　1 ft.
June—July.

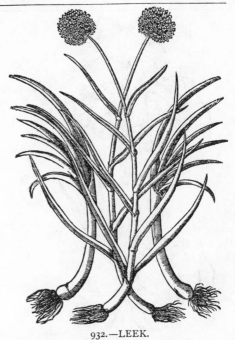

932.—LEEK.
ALLIUM PORRUM.　M. (1565) 551.
White.　2 ft.　April—May.

The leaves of the next figure have more of the
character of our leeks.

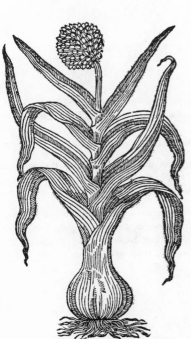

933.—"GREAT MOUNTAIN
GARLIC."　G. 142.3.

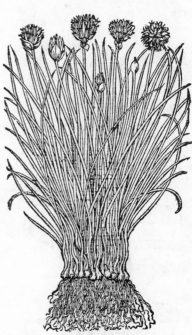

934.—CHIVES.
ALLIUM SCHŒNOPRASUM.　G. 139.1.
Flesh colour.　6 in.　May—June.

935.—ADAM'S NEEDLE.
Yucca integrissima.　J. 1543.
Greenish white.

936.—SUCCOTRINE ALOE.
Aloe perfoliata.　L. 374.1.
Red striped.　Feb.—April.

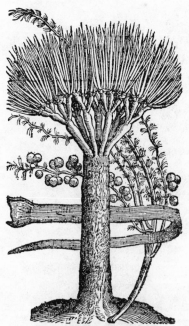

937.—DRAGON TREE.
Dracæna Draco.　J. 1523.1.

YUCCA.—The flowers are here shown about the right shape, but too large in scale and too few. The leaves consequently too small, and also too few. They usually make a great tuft with the lower ones pointing downwards.

ALOE.—A figure of this Aloe is given in the *Botanical Magazine*, No. 472. The flower spike is from 9 to 18 in. long ; the flowers themselves 1¼ in.

DRAGON TREE.—Several species are cultivated in greenhouses. This one, however, becomes a great tree. The flowers are small, the fruit baceate. Dragon's Blood is a resin derived from this tree, which grows in the Canary Isles.

MELANTHACEÆ—MEADOW SAFFRON.

Herbs with bulbous, tuberous, or fibrous roots, and regular flowers having perianths of 6 divisions. "The flowers of many are inconspicuous, and of a dull-green or yellow colour, sometimes assuming a livid hue which well bespeak the nature of their powers,"—"they are almost universally poisonous."

The Meadow Saffron or Autumn Crocus is similar in appearance to the Saffron Crocus—Crocus sativus—which is one of the IRIDACEÆ, but it has 6 instead of 3 stamens. The leaves are broadly lanceolate and wither away in the summer. Meadow Saffron is sometimes placed amongst the LILIACEÆ.

938.—WHITE HELLEBORE.
VERATRUM ALBUM. M. (1565) 1219.
Flowers light yellow. Height, 5 ft. The perianths shown erroneously with 4 instead of 6 perianths.

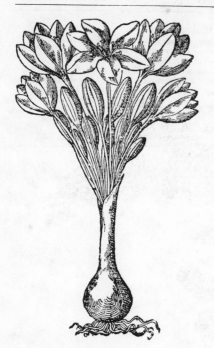

939.—HUNGARY MEAD SAFFRON.
G. 127.4.

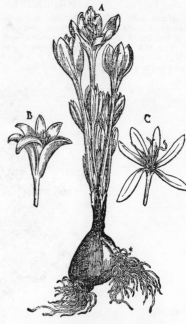

940.—MEADOW SAFFRON.
Colchicum autumnale. C. E. 845.
Purple. 4 in. Sep.—Oct.

COMMELYNACEÆ.

Perianth 6-parted, inferior, the 3 outer green the inner 3 purplish blue, petaloid. Stamens, 3—6 ; style, 1.

Tradescantia is common in gardens—growing in a dense clump.

941.—SPIDERWORT.
Tradescantia virginica. J. 49.5.
Purplish blue. 18 in. May—Oct.

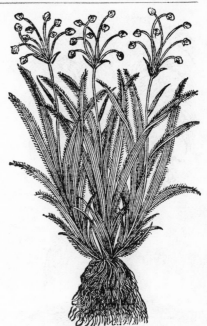

JUNCACEÆ—RUSH.

Flowers with a regular inferior 6-parted perianth, usually of glumaceous texture, sometimes petaloid. Stamens, 6; style, 1; stigmas, 3. The Hairy Wood Rush is here shown in fruit.

The COMMON RUSH, Juncus communis, has its flowers protruding in a tuft from the side of its slender stem-like leaves. Other rushes have the flowers in panicles. The inflorescence in Fig. 942 is, indeed, not accurate.

For Common Rush and flower of Luzula pilosa see p. 528.

942.—HAIRY WOOD RUSH.
LUZULA PILOSA. G. 17.1.

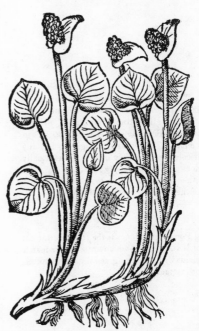

ACORACEÆ—SWEET FLAG.

Plants notable for the beautiful white spathes, which are present in some species of Calla, particularly of Calla æthiopica, commonly called the ARUM LILY, which has sagittate leaves and a graceful trumpet-shaped spathe revealing the yellow spadix within.

Calla patustris is of Northern Europe. Its spathes and leaves are both about 3 inches long.

The SWEET FLAG, Acorus Calamus, is a British plant with sword-shaped leaves, leaflike stems, a sessile spadix, but lacking the white, spathe. 2—3 ft. July.

943.—MARSH CALLA.
CALLA PATUSTRIS. Z. 724.3.
Spathe white, spadix green. July—Aug.

944.—DATE PALM.
PHŒNIX DACTYLIFERA. J. 1517.1.

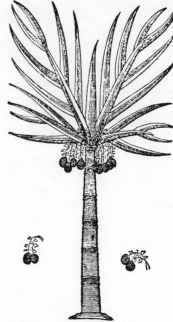

945.—COCOA-NUT PALM.
NOCOS NUCIFERA. D. App. 7.

PALMACEÆ—PALMS.

The flowers have a perianth of 6 parts
in 2 whorls, but certainly do not con-
cern us. The trunk sometimes divides
into 2. Both these trees rise to 80 feet in
height. The figures are given for any
decorative value they may have—especially
does this remark apply to Fig. 945. The
other shows the dense mass of fruit—part
of the immense spathe still remaining.

Pondweed belongs to the JUNCAGINACEÆ
or NAIADACEÆ—which are water or marsh
plants with white or green flowers in spikes
or racemes.

946.—BROAD-LEAFED PONDWEED.
POTAMOGETON NATANS. F. 651.
Scale ⅓. Stipules large and free. July.

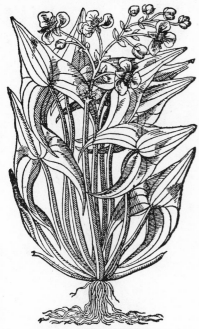

947.—ARROW-HEAD.
SAGITTARIA MINOR. G. 336.2.
White, purple spot at the base of the petals.

948.—ARROW-HEAD.
SAGITTARIA MINOR. C. E. 874.
Stalks triangular. 1–2 ft. July—Aug.

949.—WATER PLANTAIN.
ALISMA PLANTAGO. G. 337.1.
Pale purple. 3 ft. Fl., ½ in. July.

ALISMACEÆ—WATER PLANTAIN.

Floating or swamp plants, usually with a creeping perennial root. Flowers in umbels, racemes, or panicles. Sepals 3, herbaceous; petals, 3; stamens, 6; ovaries superior, several, with a style to each.

This Order, says Lindley, is to Endogens what RANUNCULACEÆ is to Exogens— having disunited carpels, and stamens from beneath them.

BUTOMACEÆ—FLOWERING RUSH.

Sometimes made part of the ALISMACEÆ. The six segments of the perianth are all petaloid. FLOWERING RUSH (Butomus umbellatus) is one of the handsomest of our native plants. Its flower head is 6 in. across, with many flowers, each 1 in. Height, 2-4 ft. Pink or white. (See Appendix, Fig. 53.)

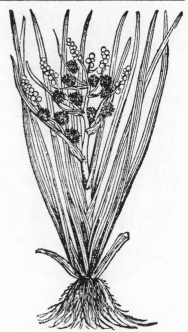

950.—CAT'S TAIL. GREAT REED-MACE.
Typha latifolia.   G. 42.
Brown, the sterile yellow.   4–6 ft.   July—Aug.

951.—BRANCHED BUR-REED.
Sparganium ramosum.   L. i. 80.1.
Stamens yellow.   2–3 ft.   July.

TYPHACEÆ—REED-MACE.

Herbaceous plants growing in marshes or ditches. Stems without nodes. Flowers unisexual, arranged on a spatheless spadix. Sepals mere scales, 3 or more, or a bundle of hairs. Petals 0, stamens 3 or 6.

The fertile flowers form the brown velvety part of the spike. The sterile flowers form a narrower continuation of the spike. Here the continuation has been overlooked, and is shown too short. Sometimes called Bull-rush, a term more properly applied to Scirpus lacustris, Fig. 958.

ARACEÆ—ARUM.

Leaves sheathing at the base, convolute in the bud, usually with branched veins.

The floral apparatus is assembled upon a spadix, within a spathe, which is of handsome form.

The upper part of the spadix, and the spathe, fall away, and the fruit remains at the summit of the stem—a cluster of berries.

HYDROCHARIDACEÆ—FROG-BIT.

Floating or water plants. Leaves with parallel veins, sometimes spiny. Flowers within a spathe. Perianth of 6 segments, the inner 3 petaloid.—Figs. 955 and 956.

952.—CUCKOW-PINT. LORDS AND LADIES.
Arum maculatum.   F. (1545) 40.
Spathe green, spadix purple, berries red.
10 in.   May.

953.—COMMON DRAGON.
ARUM DRACUNCULUS. M. 288.2.
Brown. 3 ft. June—July.

954.—COMMOM DRAGON.
ARUM DRACUNCULUS. Z. 723.2.

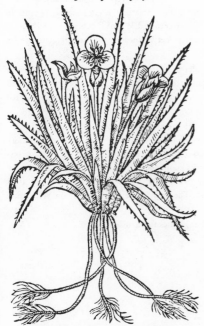

955.—WATER SOLDIER.
STRATIOTES ALOIDES. L. 375.2.
White. Scale ⅓. July.

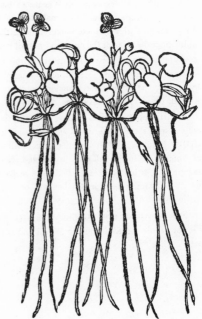

956.—FROG-BIT.
HYDROCHARIS MORSUS-RANÆ. J. 818.2.
White. Fl., 1 in. July—Aug.

[Here end the PETALOID ENDOGENS. The GLUMACEOUS ENDOGENS follow.]

GLUMACEOUS ENDOGENS.

These are the Sedges, CYPERACEÆ, and the Grasses, GRAMINACEÆ. Of these extensive Orders it has not been thought necessary to include many representatives. It is unnecessary also to go into much detail in the descriptions.

Whereas in all other flowering plants the envelope of the flower—the calyx and corolla which enclose its organs—is cup-like, even when the parts are in separate pieces, and surround the organs, here the organs are enclosed from the two opposite sides only. Lindley in beginning his descriptions of these Orders speaks of the "flowers consisting of imbricated bracts."

The simplest formation is that seen in some of the Sedges, in which the spike of flowers consists of a mass of sharp-pointed bracts each enclosing a flower. These flowers are no more than stamens and pistils.

Hence both in Sedges and Grasses there are some in which the spike is like a fox's tail. Others again have their flowers a few together upon tiny pedicels, and these again upon other slender stalks, by which they are joined into a loose panicle, as we see in Oats and some of the common Meadow Grasses.

Another kind is that in which these groups of few flowers—spikelet is the name for such a group—are so closely set against the central culm, rachis, or stalk, as to make it appear that the flowers all rise directly from the culm. Such examples we see in Wheat and Barley. So closely and regularly are the flowers packed together in some cases that the "rows" they make are easily distinguished. "Four-cornered Wheat," "Six-rowed Barley," "Two-rowed Barley"—these names indicate the regularity of the ears. Yet in these cases the spikelets are set on the opposite sides of the culm and not all round it.

CYPERACEÆ.—Stems solid, usually without joints, often triangular in section. Leaves consisting of a tubular sheath a ound the stem, not split, and a long blade thicker in substance than in grasses. Ovary often surrounded by bristles, and with 1 style branching into 2 or 3 limbs, which are not feathery. Stamens 3, with anthers fixed by their bases. Examples—Figs. 957 and 958.

GRAMINACEÆ.—Stems hollow, except at the nodes or joints, which are also wider, round in section. Leaves with a split sheath and with a ligule closing the axil where the blade leaves the stem. Ovary with 2 feathery styles. Stamens 3, with versatile anthers with curved cells. These organs are enclosed within two husks, one of which is called the palea and is double-keeled, as if it were 2 husks conjoined. The other husk is the flowering glume. It has one keel. The palea and glumes end in one or two points. More noticeable are the awns which sometimes arise from the backs of one or both of these husks. The awns make the "beard" in Wheat and Barley.

Such flowers are generally 3 or 4 together in a spikelet, with other glumes to enclose them all.

For details of the spikelets in Wheat and Barley see p. 528.

957.—GALINGALE. SWEET CYPERUS.
Cyperus longus. F. 453.
Husks brown. 2–3 ft. Bogs. July—Aug.

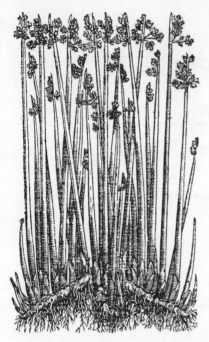

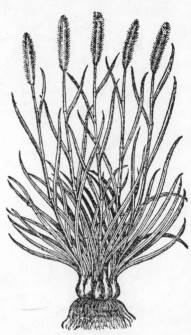

958.—BULL-RUSH [CYPERACEÆ].
SCIRPUS LACUSTRIS. G. 31.3.

959.—SLENDER FOX-TAIL GRASS.
ALOPECURUS AGRESTIS. G. 10.2.

960.—WHEAT.
TRITICUM VULGARE. M. 223.

961.—BARLEY.
HORDEUM TETRASTICHUM.
M. 224.

962.—SPELT.
TRITICUM SPELTA. M. 226.

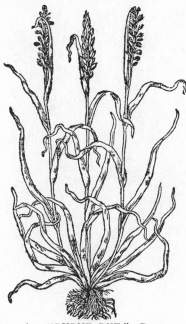

963.—"BURNT RYE." G. 70.2.

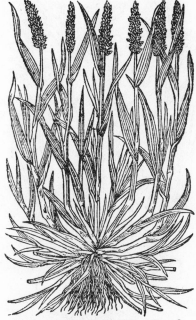

964.—PANICUM VIRIDE. G. 20.8.

965.—PANICK GRASS.
PANICUM CRUS-GALLI. G. 15.1.
2 ft. July.

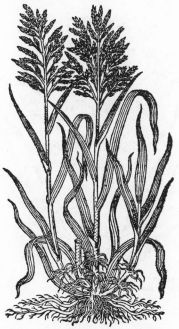

966.—REED MEADOW GRASS.
POA AQUATICA. G. 7.2.
Green. 4–6 ft. May—Aug.

967.—"FIELD BROME GRASS."
G. 69.2.

968.—PANICUM VERTICILLATUM.
G. 14.2.

969.—"PANNICK GRASS."　G. 8.1.

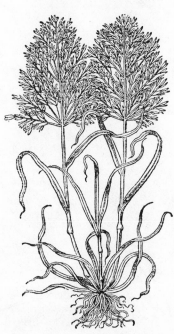

970.—TURFY HAIR-GRASS.
Aira cæspitosa.　G. 5.1.
Purple.　1–3 ft.　July.

971.—BARREN BROME GRASS.
Bromos sterilis. G. 69.1.
2 ft. June—July.

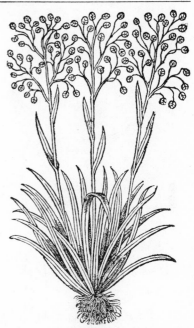

972.—QUAKING GRASS.
Briza media. J. 86.2.
Purplish. 1 ft. June.

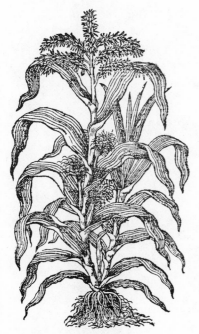

973.—MAIZE.
Zea Mays. G. 75.2.

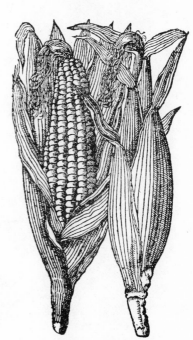

974.—EARS OF MAIZE. J. 81.3.

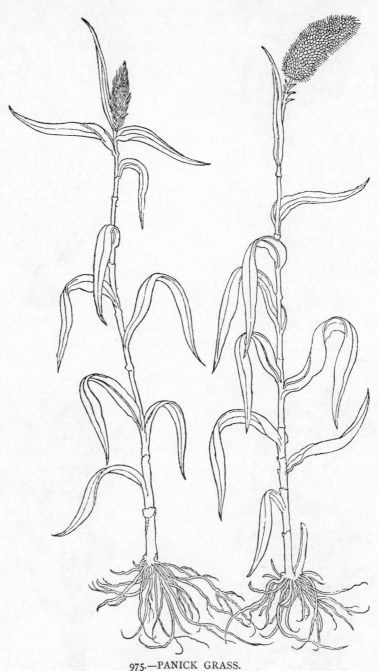

975.—PANICK GRASS.
PANICUM SATIVUM.   F. 253.

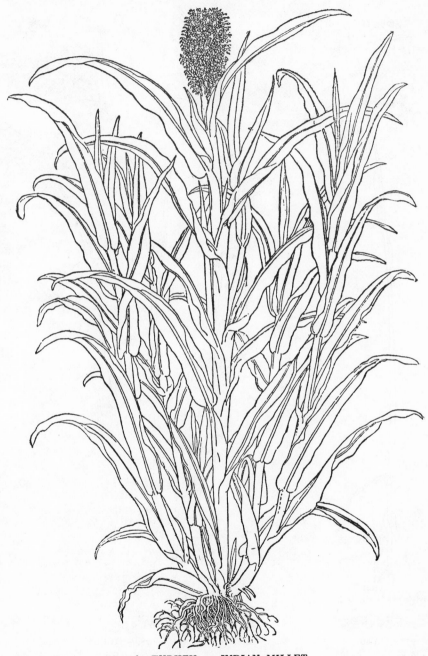

976.—TURKEY or INDIAN MILLET.
Holcus Sorghum. F. 771.
4 ft.

977.—OATS.
AVENA SATIVA. F. 185.

978.—MILLET.
PANICUM MILIACEUM. F. 411.

[Here end the ENDOGENS and all the flowering plants. These which follow are the CRYPTOGAMS, which have no proper flowers.]

CRYPTOGAMIA.

Flowerless plants, reproducing themselves by spores, which are developed without the agency of stamens and pistils, as in the flowering plants. Sufficient has been said about them on pages 39, 42, and 43. Cryptogams carry plant-life down to that simple stage where reproduction is by mere subdivision, and practically all organs are absent.

The Fungi given in Fig. 1007 are—A, Truffle, Tuber cibarium; B, Mitre Mushroom, Helvella crispa; C, Agaricus sinuatus; D, Morell, Morchella esculenta; E, Puff-ball, Lycoperdon gemmatum; F, Amanita muscaria; G, Coprinus aratus; and H, Common Mushroom, Agaricus campestris.

980.—" HANDED MOON FERN."
L. ii. 267. 1.

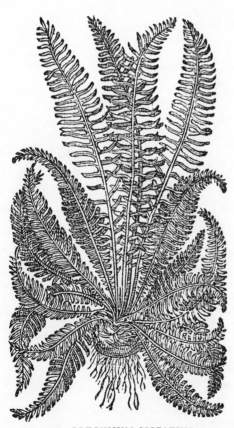

979.—BLECHNUM SPICATUM.
L. i. 815. 2.

981.—COMMON POLYPODY.
POLYPODIUM VULGARIS. G. 972.

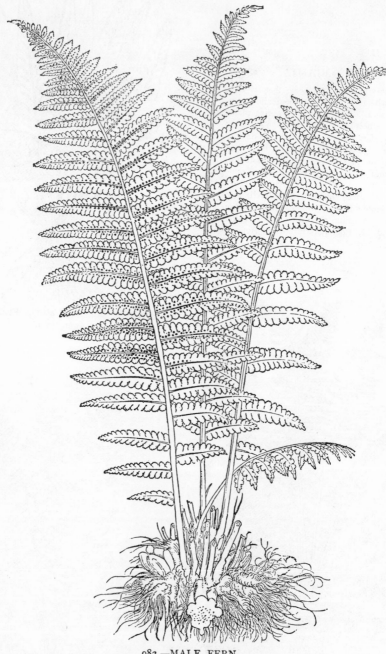

982.—MALE FERN.
Aspidium Filix-mas. F. 595.
Height, 2–5 ft. Woods and hedge-banks. June—July.
The Lady Fern, Asplenium Filix-fœmina, is similar, but the pinnules more deeply divided and serrated, and it is a more graceful fern.

480

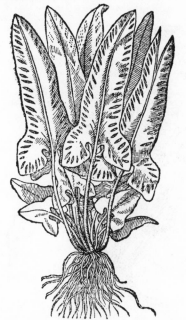

983.—"MULES FERN" OR "MOON
FERN." L. ii. 266.1.

984.—"FINGER HART'S-TONGUE."
L. ii. 266.2.

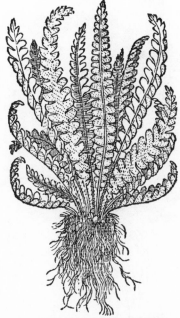

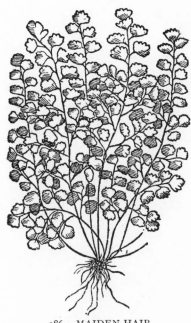

985.—COMMON SPLEENWORT.
CETERACH OFFICINARUM. L. i. 807.1.
8 in. April—Oct.

986.—MAIDEN-HAIR.
ADIANTUM CAPILLUS-VENERIS.
M. (1554) 578.1.
1 ft. May—Sep.

987.—ADIANTUM NIGRUM.
G. 975.1.

988.—HART'S-TONGUE.
Scolopendrium vulgare. Z. 897.
6 in.–3 ft. July—Aug.

989.—MAIDEN-HAIR.
G. 982.1.

990.—WALL RUE.
Asplenium Ruta-Muralia.
M. (1554) 463.
4 in. May—Sep.

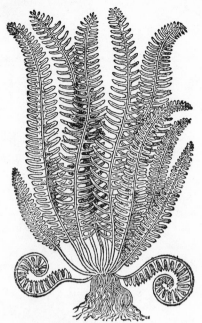

991.—LONCHITIS ASPERA. G. 978.2.

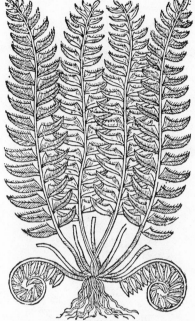

992.—POLYPODIUM LONCHITIS.
G. 979.

993.—OAK FERN.
POLYPODIUM DRYOPTERIS. G. 974.2.
1 ft. June—July.

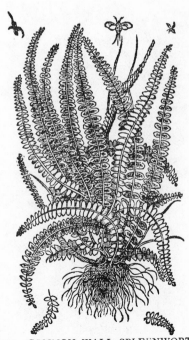

994.—COMMON WALL SPLEENWORT.
ASPLENIUM TRICHOMANES. D. 1211.

995.—MOON-WORT.
BOTRYCHIUM LUNARIA. Z. 948.2.
6 in. Aug.

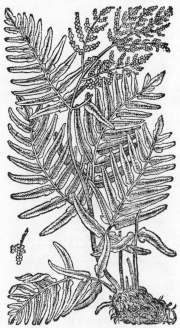

996.—FLOWERING FERN.
OSMUNDA REGALIS. D. 1225.2.
1–11 ft. Fructification in panicles.

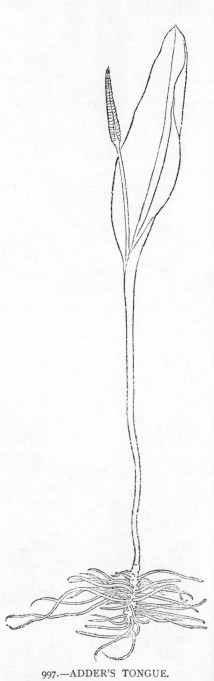

997.—ADDER'S TONGUE.
OPHIOGLOSSUM VULGATUM. F. 577.
2–12 in. May—June.

999.—HORSE-TAIL.
EQUISETUM ARVENSE. M. (1554) 459.2.

EQUISETACEÆ—HORSE-TAILS.

Leafless branched plants growing in ditches and rivers, having tubular jointed stems with the articulations separable, and surrounded by a membranous-toothed sheath. The spore cases are in terminal cones.

MARES'-TAILS are plants of somewhat similar general appearance, but have proper leaves and flowers, and belong to an order of Epigynous Calycifloral Exogens—the Haloragaceæ. Bentley says, "They are merely an imperfect form of Onagraceæ."

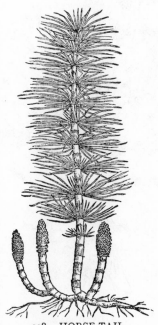

998.—HORSE-TAIL.
EQUISETUM FLUVIATILE.
M. (1554) 459.1.

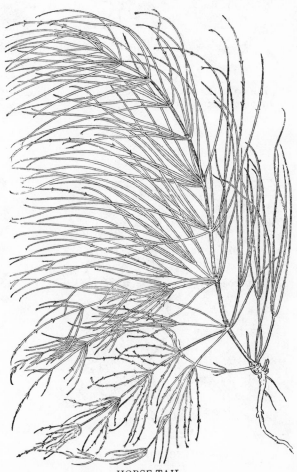

1000.—HORSE-TAIL.
EQUISETUM ARVENSE. F. 323.

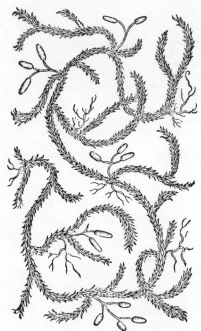

1001.—COMMON CLUB MOSS.
LYCOPODIUM CLAVATUM. C. E. 33.

1002.—"WHITE CORALINE OR SEA MOSS."
G. 1379.1.

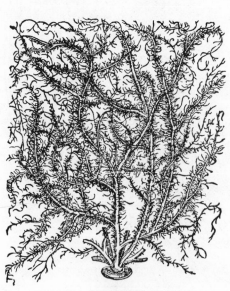

1003.—BRANCHED MOSS.
G. 1372.5.

1004.—LICHEN.
M. (1554) 462.2.

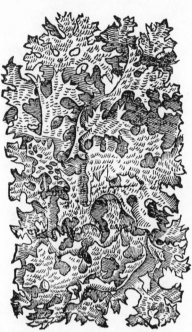

1005.—GROUND LIVERWORT.
LICHEN PULMONARIUS. J. 1565.

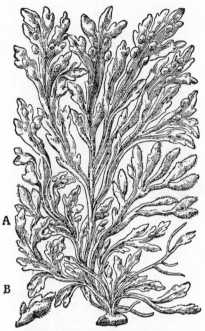

1006.—SEA OAK OR WRAKE.
FUCUS VESICULOTUS. G. 1378.3.

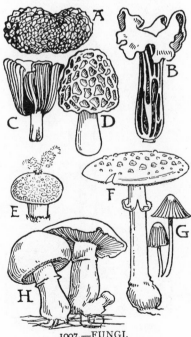

1007.—FUNGI.
(Chiefly from Lindley.)

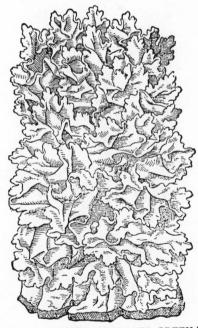

1008.—"SEA LUNGWORT OR OYSTER GREEN."
G. 1377.2.

1009.—"SAFSAF SYRORUM."
D. App. 25.

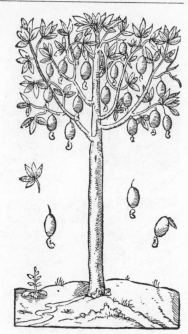

1010.—"ACAIOU."
D. 1846.2.

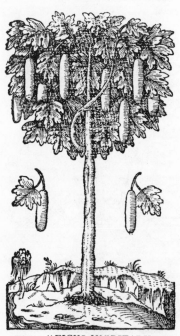

1011.—"FICUS NIGRITARUM."
D. 1842.2.

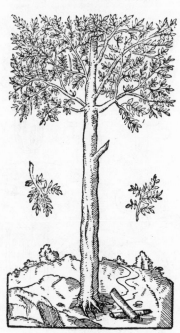

1012.—"ARBOR BRASILIA."
D. 1846.1.

(These blocks are probably of trees of Tropical America.)

488

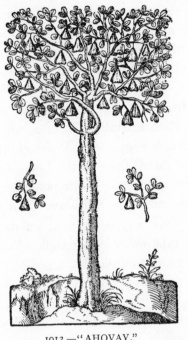

1013.—"AHOVAY."
D. 1843.1.

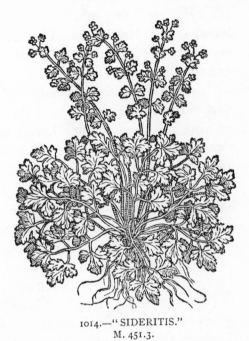

1014.—"SIDERITIS."
M. 451.3.

Fig. 1013 has been identified as Cerbera Ahovay, one of the APOCYNACEÆ, but it appears to be very untrue. The other—Sideritis—is too pretty to omit, but I cannot suggest an identification.

APPENDIX

HERE are placed certain figures which, for one reason or another, could not well be put amongst their fellows in their Natural Orders.

The first two figures represented in tone-blocks are of blocks used by Egenolph. They are given enlarged to quite twice the scale of the originals. Their bold, sweeping lines render them very valuable as suggestions for decorative treatment. If the reader will put tracing-paper over them and will follow the lines, he will find how free and swift they are. The books containing these cuts are usually crudely coloured. The colouring is not realistic, and is generally confined to a few pigments.

After these bold woodcuts follow several etchings from the *Ecphrasis* of Fabius Columna, who was a draughtsman of high ability as well as a sound and careful botanist. While his drawings must yield to those of

Fuchsius in beauty, they yet exhibit that fine sanity of observation and fidelity of rendering which lift them to a very high level. In a way, the leaves of the Hawkweeds, Figs. 10 and 11, are remarkably beautiful. In many kinds of decoration the treatment and truthfulness seen in these examples is very acceptable.

To these succeed a number of figures from the *Hortus Floridus* of Crispian van de Passe, Crispin de Passe, or Pass, according as one cares to write his name. The originals are copperplate engravings, about eight inches by four. They appeal to the gardener rather than to the botanist, and evidently exhibit an array of choice garden specimens. The plates are here reproduced in line-blocks and have lost some of their beauty. They are, however, purposely so reproduced so that they may be printed on the ordinary paper of the book, a paper which will take water-colour well. For it is the hope of the compiler that people will colour their copies.

The *Hortus Floridus* was issued in an English form as *A Garden of Flowers*. A copy is in the British Museum. It is coloured throughout, and the text consists largely of instructions for colouring the plates. Sundry verses exalt the merits, labours and expenditure of the author, and exhort the possessor not to spoil the book by blots and blurs after he has taken so much trouble. The *Hortus Floridus* concludes with a *pars altera*, an appendix, which apparently consists of engravings of a series of drawings which had come into Crispin's possession, and which he had not executed himself. Some one else, too, must have been the engraver, for the engraving is of inferior quality. Two examples are given on a later page, Figs. 56 and 57. This set of drawings was probably made by Jacques le Moyne, as has been suggested on page 9. Le Moyne's *La Clef des Champs* contains about a hundred cuts of beasts, birds and plants. There are two copies in the British Museum. The cuts were apparently intended to be coloured, as many details are added in colour.

On the opposite page we give four of the figures, and in Fig. 52 eight more.

Though one calls these cuts crude it is remarkable how true they are and how well they look when coloured. Moreover, when repeated edge to edge they make quite good patterns. This remark applies to almost any of them, but particularly to the Dasie, the Harebelles, the Cheries, and the Hazel Nuttes.

If the reader colours his copy he may gain some assistance by referring to the work of the Sowerbys and of Sydenham Edwards—the former in

English Botany and in a fine work on Fungi; the latter in the early volumes of Curtis's *Botanical Magazine*. There are some extremely well-coloured plates in Smith's *Grammar of Botany*: London, 1820. Maund's *Botanic Garden*, though elaborately done, is less successful. The reader should, of course, colour a figure from a real plant, and should not be in

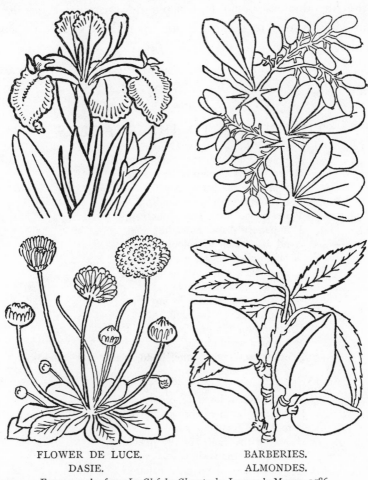

FLOWER DE LUCE.   BARBERIES.
DASIE.   ALMONDES.
Four examples from *La Clef des Champs*, by Jacques le Moyne, 1586.

a hurry to get the designs coloured—else he will only find that he has proceeded with too little experience. Let him reflect, like the possessor of Crispin, upon the author's expense, to say nothing of the publishers' or even of his own. As a rule, flat washing is a mistake. Some measure of gradation and of variation of hue there should always be. White flowers are best done with a little white, shaded with blue-grey. The head- and tail-pieces in the early part of the book will also bear colouring.

Poetry and portraits usually embellished the old herbals. The reader will perhaps not object to the inclusion of the fine portrait of Columna which is reproduced opposite, from the print on the back of the title-page of his *Ecphrasis*. It is apparently his own work.

Nor perhaps will the reader object to our completing this page with a copy of the introductory poem to Crispin de Passe's *Garden of Flowers*. The poem will be seen to be an acrostic upon the name—Crispian van de Passe, Junior.

THE BOOKE TO HIS READERS

C ome hether you that much desire,
R are flowers of dyvers landes :
I represent the same to you,
S et downe unto your hands.
P resentinge them unto your vew,
I n perfect shape and faire :
A nd also teach to colour them,
N ot missinge of a haire
V singe such couloures as requires,
A master workeman will :
N ot swarvinge thence in any case,
D eclaringe there his skill,
E ach flower his proper lineament
P resentes from top to toe :
A nd shewes both Roote, budd, blade, and stalke
S o as each one doth growe.
S paringe no paines, nor charge I have,
E ach seasons flower te passe :
I n winter, Somer, Springe and fall.
V ntill this complete was.
N ow use this same for thy delight,
I njoy it as thou wilt :
O f blotts and blurrs most carefully
R efraine, or else t'is spilt.

THOMAS WOOD.

FABIUS COLUMNA.

From the etching by himself in his *Ecphrasis*.

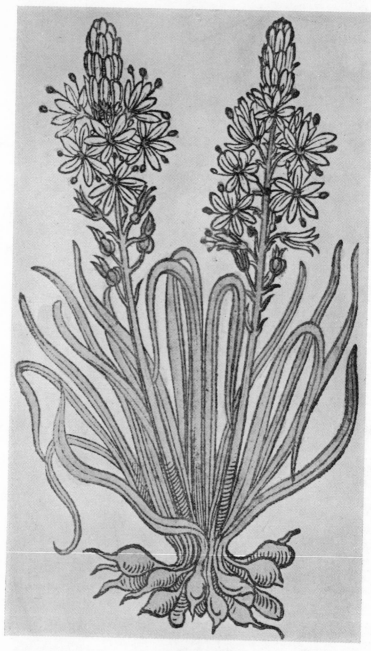

1.—KING'S SPEAR.

Asphodelus luteus. E. (*Liliaceæ.*)

2.—MYRTLE.

MYRTUS COMMUNIS. E. (*Myrtaceæ.*)

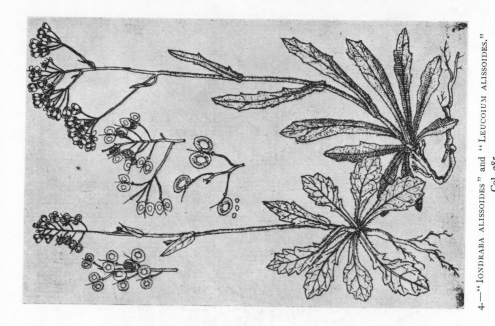

4.—"IONDRABA ALISSOIDES" and "LEUCOIUM ALISSOIDES."
Col. 285.
[Apparently species of Biscutella—Buckler Mustard.] (*Cruciferæ.*)

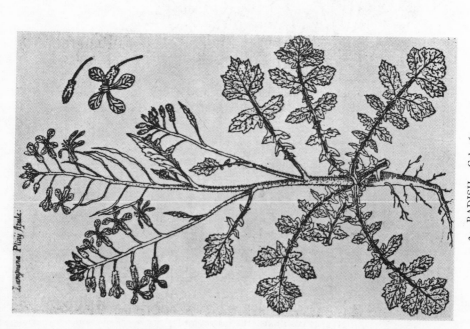

3.—RADISH. Col. 263.
A Radish with white flowers with dark lines on them.
(*Cruciferæ.*)

7.—"Hungary Mustard." Col. 276.

(*Cruciferæ.*)

6.—"Draba." Col. 272.

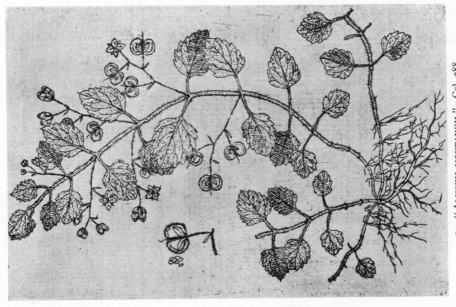

5.—"Alissum montanum." Col. 288.

[Apparently not a Crucifer; perhaps one of the Scrophulariaceæ.]

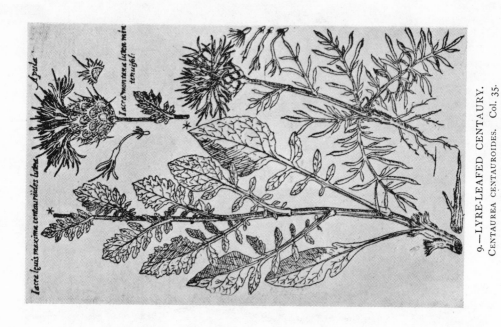

9.—LYRE-LEAFED CENTAURY.
CENTAUREA CENTAUROIDES. Col. 35.
Flowers yellow.

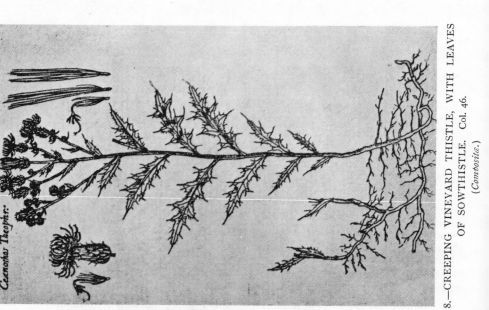

8.—CREEPING VINEYARD THISTLE, WITH LEAVES
OF SOWTHISTLE. Col. 46.
(*Combosita.*)

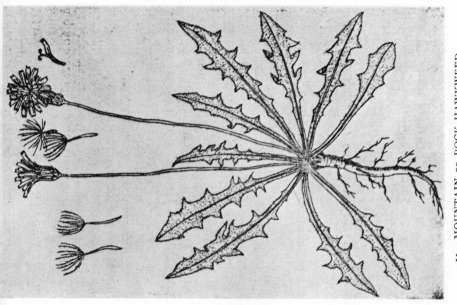

11.—MOUNTAIN or ROCK HAWKWEED.
Col. 243.

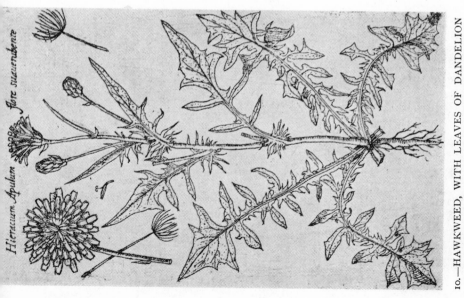

10.—HAWKWEED, WITH LEAVES OF DANDELION
AND BLUSH-COLOURED FLOWERS.   Col. 242.

(*Compositæ.*)

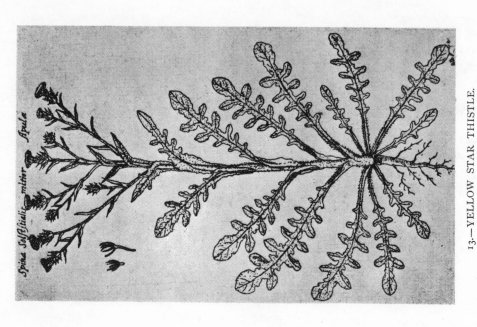

13.—YELLOW STAR THISTLE.
CENTAUREA SOLSTITIALIS. Col. 31.
If the same as the British species it is 2 ft. high and flowers June—Oct.
(*Compositæ.*)

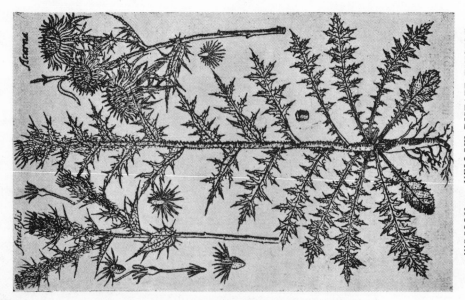

12.—WOOLLY OR YELLOW DISTAFF THISTLE.
CARTHAMUS LANATUS. Col. 23.
(Atractylis in the figure.) The Wool is shown.
Acarna is a yellow thistle. Gerarde calls it a Fish Thistle.

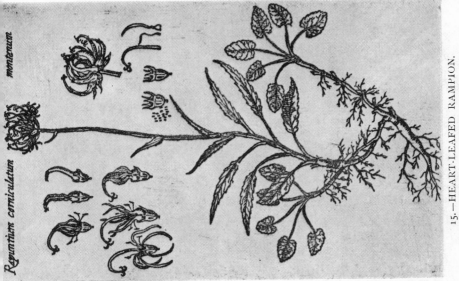

Rapuntium corniculatum *montanum*

15.—HEART-LEAFED RAMPION.
PHYTEUMA CORDATUM. Col. 224.
Blue. 6 in. July—Aug.
(*Campanulaceæ.*)

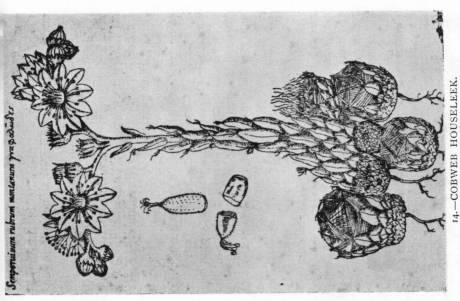

Sempervium rubrum montanum prægrandes

14.—COBWEB HOUSELEEK.
SEMPERVIVUM ARACHNOIDEUM. Col. 291.
Red. Stem and stem-leaves pink. 6 in. Fl., ¾ in. Italy. June—July.
(*Crassulaceæ.*)

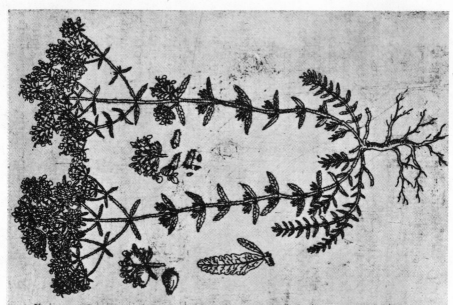

17.—HYSSOP-LEAFED GERMANDER. TEUCRIUM PSEUDO-HYSSOPUS. Col. 67. White. 18 in. June–July. (*Labiatæ*.)

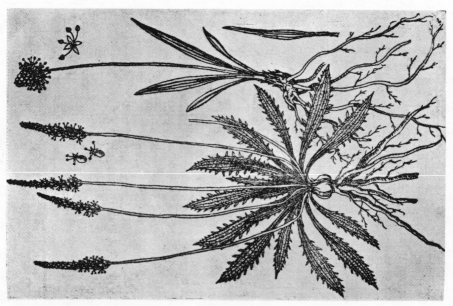

16.—HAIRY BULBOUS PLANTAIN AND THREE-NERVED PLANTAIN. Col. 259. (*Plantaginaceæ*.)

18.—THREE SPECIES OF ANEMONE. C. v. d. P. (*Ranunculaceæ.*)

19.—ANEMONE.   C. v. d. P.

20.—GREEN HELLEBORE.

HELLEBORUS VIRIDIS.   C. v. d. P.   (*Ranunculaceæ.*)

21.—CLEMATIS.   C. v. d. P.

The Fruit is erroneous.   (*Ranunculaceæ.*)

22.—LIVERWORT.

HEPATICA TRILOBA.   C. v. d. P.   (*Ranunculaceæ*.)

23.—LOVE-IN-A-MIST.

NIGELLA.   C. v. d. P.   (*Ranunculaceæ*.)

24.—POPPY.   C. v. d. P.
(*Papaveraceæ.*)

25.—CARNATIONS.
DIANTHUS.   C. v. d. P.
(*Caryophyllaceæ.*)

26.—SUPERB PINK. DIANTHUS SUPERBUS. C. v. d. P.
The flowers white or pink. (*Caryophyllaceæ.*)

27.—HOLLYHOCK. ALTHÆA ROSEA. C. v. d. P. (*Malvaceæ.*)

VARIOUSLY-COLOURED ROSE. C. v. d. P.

(*Rosaceæ.*)

28.—HUNDRED-LEAFED ROSE.

29.—SUNFLOWER. HELIANTHUS ANNUUS. C. v. d. P. (*Compositæ.*)

30.—FEVERFEW. C. v. d. P. (*Compositæ.*)

31.—PETUNIA. C. v. d. P. (*Solanaceæ.*)

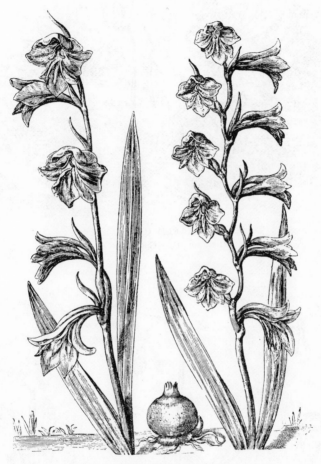

32.—GLADIOLUS. C. v. d. P. (*Iridaceæ*).

The two bracts at each flower are shown. They are curled at the edges into a tubular form. One encloses the other, which itself encloses the flower.

33.—IVY-LEAFED CYCLAMEN.
CYCLAMEN HEDERIFOLIUM.   C. v. d. P.   (*Primulaceæ.*)

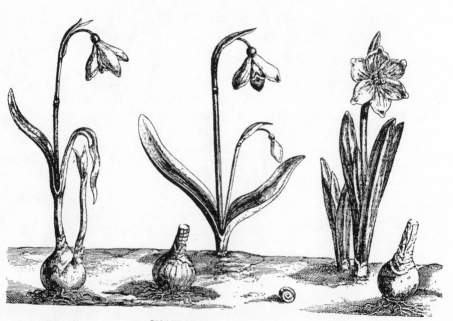

34.—SNOWDROP.   GALANTHUS NIVALIS.
White, inner segments with a green mark.   6 in.   Feb.—Mar.   (*Amaryllidaceæ.*)

35.—NARCISSUS.

NARCISSUS ALBIDUS PLENUS.   NARCISSUS INCOMPARABILIS.

C. v. d. P.   (*Amaryllidaceæ.*)

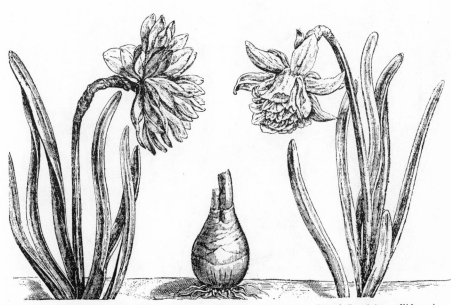

36.—DOUBLE DAFFODIL. NARCISSUS PSEUDO-NARCISSUS. C. v. d. P. (*Amaryllidaceæ.*)

37.—DAFFODIL. Narcissus Pseudo-narcissus. C. v. d. P.
Yellow. Bell with 6 lobes. 1 ft. Mar. (*Amaryllidaceæ.*)

38.—JONQUIL. Narcissus Jonquilla. C. v. d. P.
Yellow. 9 in. Ap.—May. (*Amaryllidaceæ.*)

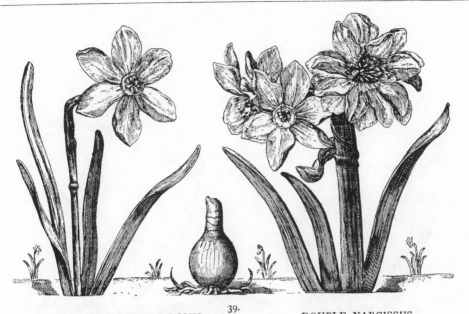

39.

PHEASANT'S EYE NARCISSUS.
NARCISSUS POETICUS. C. v. d. P.
White, the cup edged with crimson. 1 ft. May.

DOUBLE NARCISSUS.
White, middle yellow.

(*Amaryllidaceæ.*)

40.—HYACINTH. HYACINTHUS ORIENTALIS. C. v. d. P.
The left figure blue, tinged purple ; the right figure pale purple. (*Liliaceæ.*)

MOLY. Allium.

41.

GARLIC. Allium oleraceum? C. v. d. P.
Flower, purple striped.   1 ft.   July.

(*Liliaceæ.*)

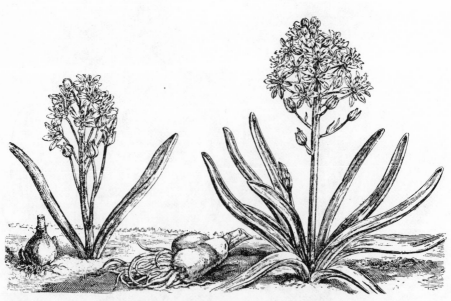

42.—SCILLA.   Flowers, blue.   C. v. d. P.   (*Liliaceæ.*)

43.—STAR OF BETHLEHEM. Ornithogalum. C. v. d. P.
Flowers, white. (*Liliaceæ.*)

44.—FRITILLARY. Fritillaria Meleagris. C. v. d. P.

The right-hand figure is the one which has purple chequered flowers and grows wild in England in moist places. Height, 9 in. Flower, 1¼ in. Mar.—May. The other figure gives the flower precisely like that of Crown Imperial, and also yellow. (*Liliaceæ.*)

45.—TULIP. TULIPA. C. v. d. P.

[We will not venture to assign more definite names to the examples on this and the next two pages. They are apparently some of the Dutch treasures— probably varieties of Tulipa Gesneriana, which Loudon says was brought from Persia in 1559. We cannot help transcribing what Loudon, after referring to its ancient popularity, says—" It is, however, like the Auricula, Pink, etc., more the poor man's flower than that of the botanists or country gentlemen."]

(*Liliaceæ*.)

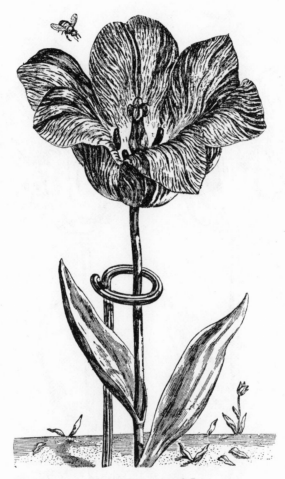

46.—TULIP. C. v. d. P.

The colours are yellow and red, sometimes white. There is a wild Tulip
(Tulipa sylvestris) in England. Its height is 1 ft., its colour yellow, and its form
not unlike the second figure on the next page. It flowers in April and May.

(*Liliaceæ.*)

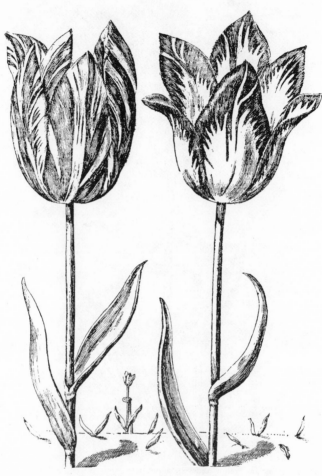

47.—TULIP.   C. v. d. P.   (*Liliaceæ.*)

48.—ORANGE LILY.

LILIUM BULBIFERUM. C. v. d. P.

Sometimes without the bulbs in the axils. Colour, orange, the central ridge outside pale green. Anthers, yellow. Style, red. Stigma, dark. There should be 6 stamens. 4 ft. June—July. (*Liliaceæ*.)

49.—ASPHODEL (see p. 458). C. v. d. P. (*Liliaceæ*)

50.—TURK'S CAP LILY. Lilium Martagon. C. v. d. P. (*Liliaceæ.*)

51.—AURICULA. C. v. d. P. (*Primulaceæ.*)

WILD HISSOPE.
BLEWBOTTLES.

HAREBELLES.
CHERIES.

HAZEL NUTTES.
EGLANTINE.

MUSKE ROSE.
WODBINE.

52.—Eight figures from *La Clef des Champs*, by Jacques le Moyne de Morgues.

53.—(*Continued opposite.*)

PINE APPLE (Ananassa sativa). CAPER (Capparis spinosa), white, stamens with pink filaments. ACACIA, flowers numerous forming balls, yellow. RHODODENDRON, white, pink, mauve or red; the lighter colours with medial red or crimson lines. EASTERN POPPY (Papaver orientalis), deep scarlet with a black spot, stamens purple. OPIUM POPPY (Papaver somniferum), white with a purple spot. BIZARRE CARNATION, white or pale yellow striped pink or purple. CANDY-TUFT (Iberis gibraltaica), flesh-colour.

54.—(*Continued opposite.*)

FRINGED WATER LILY (Villarsia nymphæoides), pale yellow, with deeper medial band. CACTUS, red. GORSE (Ulex europæus), yellow. GRASS OF PARNASSUS (Parnassia palustris), white. BEE ORCHID, sepals pinkish, the narrow petals greenish; lip, purple-brown with narrow yellow rectangular lines. CURRANT (Ribes), calyx pink, deeper towards the tips, petals small white adhering to it (a dissection shown). AZALEA, yellow or pink, usually with deeper marks. JUDAS-TREE, pink. HONEYSUCKLE (Lonicera Periclymenum), yellow, reddish outside, berries bright red. LOOSESTRIFE (Lythrum salicaria), purplish-pink, calyx crimson.

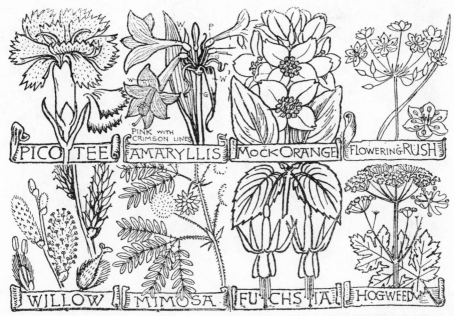

53.—(*Continued.*)

PICOTEE (Dianthus), speckled or edged with pink or purple. AMARYLLIS, pink with crimson lines. MOCK ORANGE or SYRINGA (Philadelphus), white. FLOWERING RUSH (Butomus umbellatus), deep pink, pistils crimson. WILLOW (Salix), catkins grey with yellow stamens or green ovaries. MIMOSA, yellow, fruit green. FUCHSIA, stamens pink, petals purple, calyx crimson, or petals crimson and calyx white, ovary green. HOGWEED (Heracleum Sphondylium), white tinged rose-purple.

54.—(*Continued.*)

SPEEDWELL (Veronica), blue shading into white at the middle (much enlarged). FOXGLOVE (Digitalis), white or red blending into white and yellow at base; inside is a white patch speckled with brown. LADY'S SLIPPER (Cypripedium), purple, pouch yellow. BRASSIA, white and pale green, marked with purple or brown. WILLOW HERB (Epilobium), petals mauve-pink, calyx crimson. CROCUS, yellow with brown markings. SELF-HEAL (Prunella), scales red-purple, flowers bright purple. TRUE LOVE (Paris), green, berries blue. GLADIOLUS, white or pink, marked crimson. BORAGE (Borago), bright blue, stamens and calyx purple, scales around stamens white.

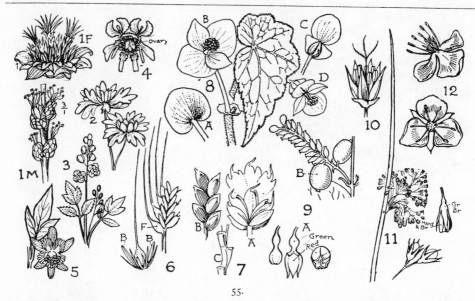

55.

1. F., SWEET CHESTNUT, female flower; 1.M., male flower (*Cupuliferæ*). 2. WINTER ACONITE, Eranthis hyemalis (*Ranunculaceæ*). Yellow. 6 in. 3. BANEBERRY, Actea spicata (*Ranunculaceæ*). White, calyx large, green. Berries black. 4. WILD MIGNONETTE, Reseda lutea. 5. SWEET MIGNONETTE (Reseda odorata) (*Resedaceæ*). 6. TWO-ROW BARLEY, a spikelet shown separately, two of the flowers, B B, being barren. F is the fertile flower. 7. WHEAT. A, a spikelet, fan-shaped, with the outline of the next above it. B, the side view of six spikelets. C, the stalk, notched to bear the spikelets alternately (*Graminaceæ*). 8. BEGONIA (*Begoniaceæ*). A, male flower in side view, unopened; C, female, ditto. B, male flower open; D, female, ditto. 9. SNOW-BERRY, Symphoria racemosa. A, flowers. B, berry (*Caprifoliaceæ*). 10. Flower of Luzula pilosa, HAIRY WOOD RUSH, and 11, flowers and details of COMMON RUSH, Juncus communis (*Juncaceæ*). 12. WATER-PLANTAIN, Alisma Plantago (*Alismaceæ*).

56.—SCARLET LYCHNIS.
LYCHNIS CHALCEDONICA. C. v. d. P. *pars altera.*

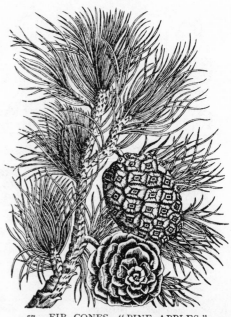

57.—FIR CONES, "PINE APPLES."
C. v. d. P. *pars altera.*

INDEX

Where a plant occurs on consecutive pages the first only is given here.   Some of the names given are synonymous with those beneath the figures.

INDEX

INDEX

INDEX

INDEX

INDEX

NOTES

LIST OF THE RARER EXOTIC PLANTS.

MOST of the figures in this book are of plants common in Britain either as natives or objects of common cultivation in gardens, or about houses. The numbers which follow are those of the figures which represent plants truly exotic, though most of them may be found in greenhouses, or hothouses.

44, 48, 49, 54, 67, 70, 102, 117, 131, 188, 189, 205 to 214, 217, 236, 237, 250, 251, 252, 263, 265, 268, 272, 273, 277, 278, 283, 287, 291 to 295, 305, 306, 307, 318, 319, 389, 390, 391, 394, 397, 398, 400, 410, 411 to 414, 423, 429, 440, 458, 470, 495, 496, 498, 500, 501, 517, 518, 523, 531, 547, 570, 571, 601, 602, 604, 608, 609, 613, 629, 630, 632 to 636, 666, 688, 698, 717, 718, 727, 728, 761, 763, 764, 775, 776, 787, 796, 798, 804, 817 to 820, 862, 863, 864, 873, 893, 894, 909, 925 to 928, 935 to 938, 943 to 945, 953, 954, 983, 1009 to 1014; Appendix—2 to 17, 53 (Caper, Acacia, and Mimosa), 54 (Judas-Tree and Brassia).

PLANTS WHICH GROW IN WATERY PLACES OR ON THE EDGE OF POOLS AND STREAMS.

23 to 27, 30, 31, 93, 148, 366, 369, 381, 443, 539, 545, 610, 615, 683, 698, 699, 702, 703, 723, 741, 743, 749, 750, 756, 764, 882, 943, 957, 980, etc. (Ferns).

PLANTS WHICH GROW IN WATER.

28, 29, 68, 69, 70, 636, 946 to 951, 955, 956, 958, 998 to 1000, Villarsia (p. 528), Butomus (p. 529), Juncus (p. 530).

ERRATA.

Page 47.—The Trilliaceæ should be amongst the Dictyogens as they are on page 433, though they are now placed by botanists amongst the Endogens as in the first reference.

A CATALOGUE OF SELECTED DOVER BOOKS
IN ALL FIELDS OF INTEREST

A CATALOGUE OF SELECTED DOVER
BOOKS IN ALL FIELDS OF INTEREST

CELESTIAL OBJECTS FOR COMMON TELESCOPES, T. W. Webb. The most used book in amateur astronomy: inestimable aid for locating and identifying nearly 4,000 celestial objects. Edited, updated by Margaret W. Mayall. 77 illustrations. Total of 645pp. 5⅜ x 8½.
20917-2, 20918-0 Pa., Two-vol. set $9.00

HISTORICAL STUDIES IN THE LANGUAGE OF CHEMISTRY, M. P. Crosland. The important part language has played in the development of chemistry from the symbolism of alchemy to the adoption of systematic nomenclature in 1892. ". . . wholeheartedly recommended,"—Science. 15 illustrations. 416pp. of text. 5⅝ x 8¼.
63702-6 Pa. $6.00

BURNHAM'S CELESTIAL HANDBOOK, Robert Burnham, Jr. Thorough, readable guide to the stars beyond our solar system. Exhaustive treatment, fully illustrated. Breakdown is alphabetical by constellation: Andromeda to Cetus in Vol. 1; Chamaeleon to Orion in Vol. 2; and Pavo to Vulpecula in Vol. 3. Hundreds of illustrations. Total of about 2000pp. 6⅛ x 9¼.
23567-X, 23568-8, 23673-0 Pa., Three-vol. set $27.85

THEORY OF WING SECTIONS: INCLUDING A SUMMARY OF AIR-FOIL DATA, Ira H. Abbott and A. E. von Doenhoff. Concise compilation of subatomic aerodynamic characteristics of modern NASA wing sections, plus description of theory. 350pp. of tables. 693pp. 5⅜ x 8½.
60586-8 Pa. $8.50

DE RE METALLICA, Georgius Agricola. Translated by Herbert C. Hoover and Lou H. Hoover. The famous Hoover translation of greatest treatise on technological chemistry, engineering, geology, mining of early modern times (1556). All 289 original woodcuts. 638pp. 6¾ x 11.
60006-8 Clothbd. $17.95

THE ORIGIN OF CONTINENTS AND OCEANS, Alfred Wegener. One of the most influential, most controversial books in science, the classic statement for continental drift. Full 1966 translation of Wegener's final (1929) version. 64 illustrations. 246pp. 5⅜ x 8½.   61708-4 Pa. $4.50

THE PRINCIPLES OF PSYCHOLOGY, William James. Famous long course complete, unabridged. Stream of thought, time perception, memory, experimental methods; great work decades ahead of its time. Still valid, useful; read in many classes. 94 figures. Total of 1391pp. 5⅜ x 8½.
20381-6, 20382-4 Pa., Two-vol. set $13.00

DRAWINGS OF WILLIAM BLAKE, William Blake. 92 plates from Book of Job, *Divine Comedy, Paradise Lost,* visionary heads, mythological figures, Laocoon, etc. Selection, introduction, commentary by Sir Geoffrey Keynes. 178pp. 8⅛ x 11. 22303-5 Pa. $4.00

ENGRAVINGS OF HOGARTH, William Hogarth. 101 of Hogarth's greatest works: *Rake's Progress, Harlot's Progress, Illustrations for Hudibras, Before and After, Beer Street and Gin Lane,* many more. Full commentary. 256pp. 11 x 13¾. 22479-1 Pa. $12.95

DAUMIER: 120 GREAT LITHOGRAPHS, Honore Daumier. Wide-ranging collection of lithographs by the greatest caricaturist of the 19th century. Concentrates on eternally popular series on lawyers, on married life, on liberated women, etc. Selection, introduction, and notes on plates by Charles F. Ramus. Total of 158pp. 9⅜ x 12¼. 23512-2 Pa. $6.00

DRAWINGS OF MUCHA, Alphonse Maria Mucha. Work reveals draftsman of highest caliber: studies for famous posters and paintings, renderings for book illustrations and ads, etc. 70 works, 9 in color; including 6 items not drawings. Introduction. List of illustrations. 72pp. 9⅜ x 12¼. (Available in U.S. only) 23672-2 Pa. $4.00

GIOVANNI BATTISTA PIRANESI: DRAWINGS IN THE PIERPONT MORGAN LIBRARY, Giovanni Battista Piranesi. For first time ever all of Morgan Library's collection, world's largest. 167 illustrations of rare Piranesi drawings—archeological, architectural, decorative and visionary. Essay, detailed list of drawings, chronology, captions. Edited by Felice Stampfle. 144pp. 9⅜ x 12¼. 23714-1 Pa. $7.50

NEW YORK ETCHINGS (1905-1949), John Sloan. All of important American artist's N.Y. life etchings. 67 works include some of his best art; also lively historical record—Greenwich Village, tenement scenes. Edited by Sloan's widow. Introduction and captions. 79pp. 8⅜ x 11¼. 23651-X Pa. $4.00

CHINESE PAINTING AND CALLIGRAPHY: A PICTORIAL SURVEY, Wan-go Weng. 69 fine examples from John M. Crawford's matchless private collection: landscapes, birds, flowers, human figures, etc., plus calligraphy. Every basic form included: hanging scrolls, handscrolls, album leaves, fans, etc. 109 illustrations. Introduction. Captions. 192pp. 8⅞ x 11¾. 23707-9 Pa. $7.95

DRAWINGS OF REMBRANDT, edited by Seymour Slive. Updated Lippmann, Hofstede de Groot edition, with definitive scholarly apparatus. All portraits, biblical sketches, landscapes, nudes, Oriental figures, classical studies, together with selection of work by followers. 550 illustrations. Total of 630pp. 9⅛ x 12¼. 21485-0, 21486-9 Pa., Two-vol. set $15.00

THE DISASTERS OF WAR, Francisco Goya. 83 etchings record horrors of Napoleonic wars in Spain and war in general. Reprint of 1st edition, plus 3 additional plates. Introduction by Philip Hofer. 97pp. 9⅜ x 8¼. 21872-4 Pa. $4.00

THE SENSE OF BEAUTY, George Santayana. Masterfully written discussion of nature of beauty, materials of beauty, form, expression; art, literature, social sciences all involved. 168pp. 5⅜ x 8½. 20238-0 Pa. $3.00

ON THE IMPROVEMENT OF THE UNDERSTANDING, Benedict Spinoza. Also contains *Ethics, Correspondence,* all in excellent R. Elwes translation. Basic works on entry to philosophy, pantheism, exchange of ideas with great contemporaries. 402pp. 5⅜ x 8½. 20250-X Pa. $4.50

THE TRAGIC SENSE OF LIFE, Miguel de Unamuno. Acknowledged masterpiece of existential literature, one of most important books of 20th century. Introduction by Madariaga. 367pp. 5⅜ x 8½.
20257-7 Pa. $4.50

THE GUIDE FOR THE PERPLEXED, Moses Maimonides. Great classic of medieval Judaism attempts to reconcile revealed religion (Pentateuch, commentaries) with Aristotelian philosophy. Important historically, still relevant in problems. Unabridged Friedlander translation. Total of 473pp. 5⅜ x 8½. 20351-4 Pa. $6.00

THE I CHING (THE BOOK OF CHANGES), translated by James Legge. Complete translation of basic text plus appendices by Confucius, and Chinese commentary of most penetrating divination manual ever prepared. Indispensable to study of early Oriental civilizations, to modern inquiring reader. 448pp. 5⅜ x 8½. 21062-6 Pa. $5.00

THE EGYPTIAN BOOK OF THE DEAD, E. A. Wallis Budge. Complete reproduction of Ani's papyrus, finest ever found. Full hieroglyphic text, interlinear transliteration, word for word translation, smooth translation. Basic work, for Egyptology, for modern study of psychic matters. Total of 533pp. 6½ x 9¼. (Available in U.S. only) 21866-X Pa. $5.95

THE GODS OF THE EGYPTIANS, E. A. Wallis Budge. Never excelled for richness, fullness: all gods, goddesses, demons, mythical figures of Ancient Egypt; their legends, rites, incarnations, variations, powers, etc. Many hieroglyphic texts cited. Over 225 illustrations, plus 6 color plates. Total of 988pp. 6⅛ x 9¼. (Available in U.S. only)
22055-9, 22056-7 Pa., Two-vol. set $16.00

THE STANDARD BOOK OF QUILT MAKING AND COLLECTING, Marguerite Ickis. Full information, full-sized patterns for making 46 traditional quilts, also 150 other patterns. Quilted cloths, lame, satin quilts, etc. 483 illustrations. 273pp. 6⅞ x 9⅝. 20582-7 Pa. $4.95

CORAL GARDENS AND THEIR MAGIC, Bronsilaw Malinowski. Classic study of the methods of tilling the soil and of agricultural rites in the Trobriand Islands of Melanesia. Author is one of the most important figures in the field of modern social anthropology. 143 illustrations. Indexes. Total of 911pp. of text. 5⅝ x 8¼. (Available in U.S. only)
23597-1 Pa. $12.95

THE PHILOSOPHY OF HISTORY, Georg W. Hegel. Great classic of Western thought develops concept that history is not chance but a rational process, the evolution of freedom. 457pp. 5⅜ x 8½.   20112-0 Pa. $4.50

LANGUAGE, TRUTH AND LOGIC, Alfred J. Ayer. Famous, clear introduction to Vienna, Cambridge schools of Logical Positivism. Role of philosophy, elimination of metaphysics, nature of analysis, etc. 160pp. 5⅜ x 8½. (Available in U.S. only)   20010-8 Pa. $2.00

A PREFACE TO LOGIC, Morris R. Cohen. Great City College teacher in renowned, easily followed exposition of formal logic, probability, values, logic and world order and similar topics; no previous background needed. 209pp. 5⅜ x 8½.   23517-3 Pa. $3.50

REASON AND NATURE, Morris R. Cohen. Brilliant analysis of reason and its multitudinous ramifications by charismatic teacher. Interdisciplinary, synthesizing work widely praised when it first appeared in 1931. Second (1953) edition. Indexes. 496pp. 5⅜ x 8½.   23633-1 Pa. $6.50

AN ESSAY CONCERNING HUMAN UNDERSTANDING, John Locke. The only complete edition of enormously important classic, with authoritative editorial material by A. C. Fraser. Total of 1176pp. 5⅜ x 8½.
20530-4, 20531-2 Pa., Two-vol. set $16.00

HANDBOOK OF MATHEMATICAL FUNCTIONS WITH FORMULAS, GRAPHS, AND MATHEMATICAL TABLES, edited by Milton Abramowitz and Irene A. Stegun. Vast compendium: 29 sets of tables, some to as high as 20 places. 1,046pp. 8 x 10½.   61272-4 Pa. $14.95

MATHEMATICS FOR THE PHYSICAL SCIENCES, Herbert S. Wilf. Highly acclaimed work offers clear presentations of vector spaces and matrices, orthogonal functions, roots of polynomial equations, conformal mapping, calculus of variations, etc. Knowledge of theory of functions of real and complex variables is assumed. Exercises and solutions. Index. 284pp. 5⅝ x 8¼.   63635-6 Pa. $5.00

THE PRINCIPLE OF RELATIVITY, Albert Einstein et al. Eleven most important original papers on special and general theories. Seven by Einstein, two by Lorentz, one each by Minkowski and Weyl. All translated, unabridged. 216pp. 5⅜ x 8½.   60081-5 Pa. $3.50

THERMODYNAMICS, Enrico Fermi. A classic of modern science. Clear, organized treatment of systems, first and second laws, entropy, thermodynamic potentials, gaseous reactions, dilute solutions, entropy constant. No math beyond calculus required. Problems. 160pp. 5⅜ x 8½.
60361-X Pa. $3.00

ELEMENTARY MECHANICS OF FLUIDS, Hunter Rouse. Classic undergraduate text widely considered to be far better than many later books. Ranges from fluid velocity and acceleration to role of compressibility in fluid motion. Numerous examples, questions, problems. 224 illustrations. 376pp. 5⅝ x 8¼.   63699-2 Pa. $5.00

THE COMPLETE BOOK OF DOLL MAKING AND COLLECTING, Catherine Christopher. Instructions, patterns for dozens of dolls, from rag doll on up to elaborate, historically accurate figures. Mould faces, sew clothing, make doll houses, etc. Also collecting information. Many illustrations. 288pp. 6 x 9. 22066-4 Pa. $4.50

THE DAGUERREOTYPE IN AMERICA, Beaumont Newhall. Wonderful portraits, 1850's townscapes landscapes; full text plus 104 photographs. The basic book. Enlarged 1976 edition. 272pp. 8¼ x 11¼. 23322-7 Pa. $7.95

CRAFTSMAN HOMES, Gustav Stickley. 296 architectural drawings, floor plans, and photographs illustrate 40 different kinds of "Mission-style" homes from The Craftsman (1901-16), voice of American style of simplicity and organic harmony. Thorough coverage of Craftsman idea in text and picture, now collector's item. 224pp. 8⅛ x 11. 23791-5 Pa. $6.00

PEWTER-WORKING: INSTRUCTIONS AND PROJECTS, Burl N. Osborn. & Gordon O. Wilber. Introduction to pewter-working for amateur craftsman. History and characteristics of pewter; tools, materials, step-by-step instructions. Photos, line drawings, diagrams. Total of 160pp. 7⅞ x 10¾. 23786-9 Pa. $3.50

THE GREAT CHICAGO FIRE, edited by David Lowe. 10 dramatic, eye-witness accounts of the 1871 disaster, including one of the aftermath and rebuilding, plus 70 contemporary photographs and illustrations of the ruins—courthouse, Palmer House, Great Central Depot, etc. Introduction by David Lowe. 87pp. 8¼ x 11. 23771-0 Pa. $4.00

SILHOUETTES: A PICTORIAL ARCHIVE OF VARIED ILLUSTRATIONS, edited by Carol Belanger Grafton. Over 600 silhouettes from the 18th to 20th centuries include profiles and full figures of men and women, children, birds and animals, groups and scenes, nature, ships, an alphabet. Dozens of uses for commercial artists and craftspeople. 144pp. 8⅜ x 11¼. 23781-8 Pa. $4.50

ANIMALS: 1,419 COPYRIGHT-FREE ILLUSTRATIONS OF MAMMALS, BIRDS, FISH, INSECTS, ETC., edited by Jim Harter. Clear wood engravings present, in extremely lifelike poses, over 1,000 species of animals. One of the most extensive copyright-free pictorial sourcebooks of its kind. Captions. Index. 284pp. 9 x 12. 23766-4 Pa. $8.95

INDIAN DESIGNS FROM ANCIENT ECUADOR, Frederick W. Shaffer. 282 original designs by pre-Columbian Indians of Ecuador (500-1500 A.D.). Designs include people, mammals, birds, reptiles, fish, plants, heads, geometric designs. Use as is or alter for advertising, textiles, leathercraft, etc. Introduction. 95pp. 8¾ x 11¼. 23764-8 Pa. $3.50

SZIGETI ON THE VIOLIN, Joseph Szigeti. Genial, loosely structured tour by premier violinist, featuring a pleasant mixture of reminiscenes, insights into great music and musicians, innumerable tips for practicing violinists. 385 musical passages. 256pp. 5⅝ x 8¼. 23763-X Pa. $4.00

TONE POEMS, SERIES II: TILL EULENSPIEGELS LUSTIGE STREICHE, ALSO SPRACH ZARATHUSTRA, AND EIN HELDEN-LEBEN, Richard Strauss. Three important orchestral works, including very popular *Till Eulenspiegel's Marry Pranks,* reproduced in full score from original editions. Study score. 315pp. 9⅜ x 12¼. (Available in U.S. only)
23755-9 Pa. $8.95

TONE POEMS, SERIES I: DON JUAN, TOD UND VERKLARUNG AND DON QUIXOTE, Richard Strauss. Three of the most often performed and recorded works in entire orchestral repertoire, reproduced in full score from original editions. Study score. 286pp. 9⅜ x 12¼. (Available in U.S. only)
23754-0 Pa. $7.50

11 LATE STRING QUARTETS, Franz Joseph Haydn. The form which Haydn defined and "brought to perfection." *(Grove's).* 11 string quartets in complete score, his last and his best. The first in a projected series of the complete Haydn string quartets. Reliable modern Eulenberg edition, otherwise difficult to obtain. 320pp. 8⅜ x 11¼. (Available in U.S. only)
23753-2 Pa. $7.50

FOURTH, FIFTH AND SIXTH SYMPHONIES IN FULL SCORE, Peter Ilyitch Tchaikovsky. Complete orchestral scores of Symphony No. 4 in F Minor, Op. 36; Symphony No. 5 in E Minor, Op. 64; Symphony No. 6 in B Minor, "Pathetique," Op. 74. Bretikopf & Hartel eds. Study score. 480pp. 9⅜ x 12¼. 23861-X Pa. $10.95

THE MARRIAGE OF FIGARO: COMPLETE SCORE, Wolfgang A. Mozart. Finest comic opera ever written. Full score, not to be confused with piano renderings. Peters edition. Study score. 448pp. 9⅜ x 12¼. (Available in U.S. only)
23751-6 Pa. $11.95

"IMAGE" ON THE ART AND EVOLUTION OF THE FILM, edited by Marshall Deutelbaum. Pioneering book brings together for first time 38 groundbreaking articles on early silent films from *Image* and 263 illustrations newly shot from rare prints in the collection of the International Museum of Photography. A landmark work. Index. 256pp. 8¼ x 11.
23777-X Pa. $8.95

AROUND-THE-WORLD COOKY BOOK, Lois Lintner Sumption and Marguerite Lintner Ashbrook. 373 cooky and frosting recipes from 28 countries (America, Austria, China, Russia, Italy, etc.) include Viennese kisses, rice wafers, London strips, lady fingers, hony, sugar spice, maple cookies, etc. Clear instructions. All tested. 38 drawings. 182pp. 5⅜ x 8.
23802-4 Pa. $2.50

THE ART NOUVEAU STYLE, edited by Roberta Waddell. 579 rare photographs, not available elsewhere, of works in jewelry, metalwork, glass, ceramics, textiles, architecture and furniture by 175 artists—Mucha, Seguy, Lalique, Tiffany, Gaudin, Hohlwein, Saarinen, and many others. 288pp. 8⅜ x 11¼. 23515-7 Pa. $6.95

THE AMERICAN SENATOR, Anthony Trollope. Little known, long unavailable Trollope novel on a grand scale. Here are humorous comment on American vs. English culture, and stunning portrayal of a heroine/villainess. Superb evocation of Victorian village life. 561pp. 5⅜ x 8½.
23801-6 Pa. $6.00

WAS IT MURDER? James Hilton. The author of *Lost Horizon* and *Goodbye, Mr. Chips* wrote one detective novel (under a pen-name) which was quickly forgotten and virtually lost, even at the height of Hilton's fame. This edition brings it back—a finely crafted public school puzzle resplendent with Hilton's stylish atmosphere. A thoroughly English thriller by the creator of Shangri-la. 252pp. 5⅜ x 8. (Available in U.S. only)
23774-5 Pa. $3.00

CENTRAL PARK: A PHOTOGRAPHIC GUIDE, Victor Laredo and Henry Hope Reed. 121 superb photographs show dramatic views of Central Park: Bethesda Fountain, Cleopatra's Needle, Sheep Meadow, the Blockhouse, plus people engaged in many park activities: ice skating, bike riding, etc. Captions by former Curator of Central Park, Henry Hope Reed, provide historical view, changes, etc. Also photos of N.Y. landmarks on park's periphery. 96pp. 8½ x 11.
23750-8 Pa. $4.50

NANTUCKET IN THE NINETEENTH CENTURY, Clay Lancaster. 180 rare photographs, stereographs, maps, drawings and floor plans recreate unique American island society. Authentic scenes of shipwreck, lighthouses, streets, homes are arranged in geographic sequence to provide walking-tour guide to old Nantucket existing today. Introduction, captions. 160pp. 8⅞ x 11¾.
23747-8 Pa. $6.95

STONE AND MAN: A PHOTOGRAPHIC EXPLORATION, Andreas Feininger. 106 photographs by *Life* photographer Feininger portray man's deep passion for stone through the ages. Stonehenge-like megaliths, fortified towns, sculpted marble and crumbling tenements show textures, beauties, fascination. 128pp. 9¼ x 10¾.
23756-7 Pa. $5.95

CIRCLES, A MATHEMATICAL VIEW, D. Pedoe. Fundamental aspects of college geometry, non-Euclidean geometry, and other branches of mathematics: representing circle by point. Poincare model, isoperimetric property, etc. Stimulating recreational reading. 66 figures. 96pp. 5⅝ x 8¼.
63698-4 Pa. $2.75

THE DISCOVERY OF NEPTUNE, Morton Grosser. Dramatic scientific history of the investigations leading up to the actual discovery of the eighth planet of our solar system. Lucid, well-researched book by well-known historian of science. 172pp. 5⅜ x 8½.
23726-5 Pa. $3.00

THE DEVIL'S DICTIONARY. Ambrose Bierce. Barbed, bitter, brilliant witticisms in the form of a dictionary. Best, most ferocious satire America has produced. 145pp. 5⅜ x 8½.
20487-1 Pa. $2.00

HISTORY OF BACTERIOLOGY, William Bulloch. The only comprehensive history of bacteriology from the beginnings through the 19th century. Special emphasis is given to biography-Leeuwenhoek, etc. Brief accounts of 350 bacteriologists form a separate section. No clearer, fuller study, suitable to scientists and general readers, has yet been written. 52 illustrations. 448pp. 5⅝ x 8¼. 23761-3 Pa. $6.50

THE COMPLETE NONSENSE OF EDWARD LEAR, Edward Lear. All nonsense limericks, zany alphabets, Owl and Pussycat, songs, nonsense botany, etc., illustrated by Lear. Total of 321pp. 5⅜ x 8½. (Available in U.S. only) 20167-8 Pa. $3.95

INGENIOUS MATHEMATICAL PROBLEMS AND METHODS, Louis A. Graham. Sophisticated material from Graham *Dial*, applied and pure; stresses solution methods. Logic, number theory, networks, inversions, etc. 237pp. 5⅜ x 8½. 20545-2 Pa. $4.50

BEST MATHEMATICAL PUZZLES OF SAM LOYD, edited by Martin Gardner. Bizarre, original, whimsical puzzles by America's greatest puzzler. From fabulously rare *Cyclopedia,* including famous 14-15 puzzles, the Horse of a Different Color, 115 more. Elementary math. 150 illustrations. 167pp. 5⅜ x 8½. 20498-7 Pa. $2.75

THE BASIS OF COMBINATION IN CHESS, J. du Mont. Easy-to-follow, instructive book on elements of combination play, with chapters on each piece and every powerful combination team—two knights, bishop and knight, rook and bishop, etc. 250 diagrams. 218pp. 5⅜ x 8½. (Available in U.S. only) 23644-7 Pa. $3.50

MODERN CHESS STRATEGY, Ludek Pachman. The use of the queen, the active king, exchanges, pawn play, the center, weak squares, etc. Section on rook alone worth price of the book. Stress on the moderns. Often considered the most important book on strategy. 314pp. 5⅜ x 8½. 20290-9 Pa. $4.50

LASKER'S MANUAL OF CHESS, Dr. Emanuel Lasker. Great world champion offers very thorough coverage of all aspects of chess. Combinations, position play, openings, end game, aesthetics of chess, philosophy of struggle, much more. Filled with analyzed games. 390pp. 5⅜ x 8½. 20640-8 Pa. $5.00

500 MASTER GAMES OF CHESS, S. Tartakower, J. du Mont. Vast collection of great chess games from 1798-1938, with much material nowhere else readily available. Fully annotated, arranged by opening for easier study. 664pp. 5⅜ x 8½. 23208-5 Pa. $7.50

A GUIDE TO CHESS ENDINGS, Dr. Max Euwe, David Hooper. One of the finest modern works on chess endings. Thorough analysis of the most frequently encountered endings by former world champion. 331 examples, each with diagram. 248pp. 5⅜ x 8½. 23332-4 Pa. $3.75

SECOND PIATIGORSKY CUP, edited by Isaac Kashdan. One of the greatest tournament books ever produced in the English language. All 90 games of the 1966 tournament, annotated by players, most annotated by both players. Features Petrosian, Spassky, Fischer, Larsen, six others. 228pp. 5⅜ x 8½. 23572-6 Pa. $3.50

ENCYCLOPEDIA OF CARD TRICKS, revised and edited by Jean Hugard. How to perform over 600 card tricks, devised by the world's greatest magicians: impromptus, spelling tricks, key cards, using special packs, much, much more. Additional chapter on card technique. 66 illustrations. 402pp. 5⅜ x 8½. (Available in U.S. only) 21252-1 Pa. $3.95

MAGIC: STAGE ILLUSIONS, SPECIAL EFFECTS AND TRICK PHO-TOGRAPHY, Albert A. Hopkins, Henry R. Evans. One of the great classics; fullest, most authoritive explanation of vanishing lady, levitations, scores of other great stage effects. Also small magic, automata, stunts. 446 illus-trations. 556pp. 5⅜ x 8½. 23344-8 Pa. $6.95

THE SECRETS OF HOUDINI, J. C. Cannell. Classic study of Houdini's incredible magic, exposing closely-kept professional secrets and revealing, in general terms, the whole art of stage magic. 67 illustrations. 279pp. 5⅜ x 8½. 22913-0 Pa. $3.00

HOFFMANN'S MODERN MAGIC, Professor Hoffmann. One of the best, and best-known, magicians' manuals of the past century. Hundreds of tricks from card tricks and simple sleight of hand to elaborate illusions involving construction of complicated machinery. 332 illustrations. 563pp. 5⅜ x 8½. 23623-4 Pa. $6.00

MADAME PRUNIER'S FISH COOKERY BOOK, Mme. S. B. Prunier. More than 1000 recipes from world famous Prunier's of Paris and London, specially adapted here for American kitchen. Grilled tournedos with anchovy butter, Lobster a la Bordelaise, Prunier's prized desserts, more. Glossary. 340pp. 5⅜ x 8½. (Available in U.S. only) 22679-4 Pa. $3.00

FRENCH COUNTRY COOKING FOR AMERICANS, Louis Diat. 500 easy-to-make, authentic provincial recipes compiled by former head chef at New York's Fitz-Carlton Hotel: onion soup, lamb stew, potato pie, more. 309pp. 5⅜ x 8½. 23665-X Pa. $3.95

SAUCES, FRENCH AND FAMOUS, Louis Diat. Complete book gives over 200 specific recipes: bechamel, Bordelaise, hollandaise, Cumberland, apri-cot, etc. Author was one of this century's finest chefs, originator of vichyssoise and many other dishes. Index. 156pp. 5⅜ x 8. 23663-3 Pa. $2.50

TOLL HOUSE TRIED AND TRUE RECIPES, Ruth Graves Wakefield. Authentic recipes from the famous Mass. restaurant: popovers, veal and ham loaf, Toll House baked beans, chocolate cake crumb pudding, much more. Many helpful hints. Nearly 700 recipes. Index. 376pp. 5⅜ x 8½. 23560-2 Pa. $4.50

"OSCAR" OF THE WALDORF'S COOKBOOK, Oscar Tschirky. Famous American chef reveals 3455 recipes that made Waldorf great; cream of French, German, American cooking, in all categories. Full instructions, easy home use. 1896 edition. 907pp. 6⅝ x 9⅜. 20790-0 Clothbd. $15.00

COOKING WITH BEER, Carole Fahy. Beer has as superb an effect on food as wine, and at fraction of cost. Over 250 recipes for appetizers, soups, main dishes, desserts, breads, etc. Index. 144pp. 5⅜ x 8½. (Available in U.S. only) 23661-7 Pa. $2.50

STEWS AND RAGOUTS, Kay Shaw Nelson. This international cookbook offers wide range of 108 recipes perfect for everyday, special occasions, meals-in-themselves, main dishes. Economical, nutritious, easy-to-prepare: goulash, Irish stew, boeuf bourguignon, etc. Index. 134pp. 5⅜ x 8½. 23662-5 Pa. $2.50

DELICIOUS MAIN COURSE DISHES, Marian Tracy. Main courses are the most important part of any meal. These 200 nutritious, economical recipes from around the world make every meal a delight. "I . . . have found it so useful in my own household,"—N.Y. Times. Index. 219pp. 5⅜ x 8½. 23664-1 Pa. $3.00

FIVE ACRES AND INDEPENDENCE, Maurice G. Kains. Great back-to-the-land classic explains basics of self-sufficient farming: economics, plants, crops, animals, orchards, soils, land selection, host of other necessary things. Do not confuse with skimpy faddist literature; Kains was one of America's greatest agriculturalists. 95 illustrations. 397pp. 5⅜ x 8½. 20974-1 Pa. $3.95

A PRACTICAL GUIDE FOR THE BEGINNING FARMER, Herbert Jacobs. Basic, extremely useful first book for anyone thinking about moving to the country and starting a farm. Simpler than Kains, with greater emphasis on country living in general. 246pp. 5⅜ x 8½. 23675-7 Pa. $3.50

A GARDEN OF PLEASANT FLOWERS (PARADISI IN SOLE: PARADISUS TERRESTRIS), John Parkinson. Complete, unabridged reprint of first (1629) edition of earliest great English book on gardens and gardening. More than 1000 plants & flowers of Elizabethan, Jacobean garden fully described, most with woodcut illustrations. Botanically very reliable, a "speaking garden" of exceeding charm. 812 illustrations. 628pp. 8½ x 12¼. 23392-8 Clothbd. $25.00

ACKERMANN'S COSTUME PLATES, Rudolph Ackermann. Selection of 96 plates from the Repository of Arts, best published source of costume for English fashion during the early 19th century. 12 plates also in color. Captions, glossary and introduction by editor Stella Blum. Total of 120pp. 8⅜ x 11¼. 23690-0 Pa. $4.50

THE CURVES OF LIFE, Theodore A. Cook. Examination of shells, leaves, horns, human body, art, etc., in *"the* classic reference on how the golden ratio applies to spirals and helices in nature . . . . "—Martin Gardner. 426 illustrations. Total of 512pp. 5⅜ x 8½.   23701-X Pa. $5.95

AN ILLUSTRATED FLORA OF THE NORTHERN UNITED STATES AND CANADA, Nathaniel L. Britton, Addison Brown. Encyclopedic work covers 4666 species, ferns on up. Everything. Full botanical information, illustration for each. This earlier edition is preferred by many to more recent revisions. 1913 edition. Over 4000 illustrations, total of 2087pp. 6⅛ x 9¼.   22642-5, 22643-3, 22644-1 Pa., Three-vol. set $25.50

MANUAL OF THE GRASSES OF THE UNITED STATES, A. S. Hitchcock, U.S. Dept. of Agriculture. The basic study of American grasses, both indigenous and escapes, cultivated and wild. Over 1400 species. Full descriptions, information. Over 1100 maps, illustrations. Total of 1051pp. 5⅜ x 8½.   22717-0, 22718-9 Pa., Two-vol. set $15.00

THE CACTACEAE,, Nathaniel L. Britton, John N. Rose. Exhaustive, definitive. Every cactus in the world. Full botanical descriptions. Thorough statement of nomenclatures, habitat, detailed finding keys. The one book needed by every cactus enthusiast. Over 1275 illustrations. Total of 1080pp. 8 x 10¼. ·   21191-6, 21192-4 Clothbd., Two-vol. set $35.00

AMERICAN MEDICINAL PLANTS, Charles F. Millspaugh. Full descriptions, 180 plants covered: history; physical description; methods of preparation with all chemical constituents extracted; all claimed curative or adverse effects. 180 full-page plates. Classification table. 804pp. 6½ x 9¼.
                                    23034-1 Pa. $12.95

A MODERN HERBAL, Margaret Grieve. Much the fullest, most exact, most useful compilation of herbal material. Gigantic alphabetical encyclopedia, from aconite to zedoary, gives botanical information, medical properties, folklore, economic uses, and much else. Indispensable to serious reader. 161 illustrations. 888pp. 6½ x 9¼. (Available in U.S. only)
                        22798-7, 22799-5 Pa., Two-vol. set $13.00

THE HERBAL or GENERAL HISTORY OF PLANTS, John Gerard. The 1633 edition revised and enlarged by Thomas Johnson. Containing almost 2850 plant descriptions and 2705 superb illustrations, Gerard's *Herbal* is a monumental work, the book all modern English herbals are derived from, the one herbal every serious enthusiast should have in its entirety. Original editions are worth perhaps $750. 1678pp. 8½ x 12¼.
                                    23147-X Clothbd. $50.00

MANUAL OF THE TREES OF NORTH AMERICA, Charles S. Sargent. The basic survey of every native tree and tree-like shrub, 717 species in all. Extremely full descriptions, information on habitat, growth, locales, economics, etc. Necessary to every serious tree lover. Over 100 finding keys. 783 illustrations. Total of 986pp. 5⅜ x 8½.
                        20277-1, 20278-X Pa., Two-vol. set $11.00

AMERICAN BIRD ENGRAVINGS, Alexander Wilson et al. All 76 plates. from Wilson's *American Ornithology* (1808-14), most important ornithological work before Audubon, plus 27 plates from the supplement (1825-33) by Charles Bonaparte. Over 250 birds portrayed. 8 plates also reproduced in full color. 111pp. 9⅜ x 12½. 23195-X Pa. $6.00

CRUICKSHANK'S PHOTOGRAPHS OF BIRDS OF AMERICA, Allan D. Cruickshank. Great ornithologist, photographer presents 177 closeups, groupings, panoramas, flightings, etc., of about 150 different birds. Expanded *Wings in the Wilderness*. Introduction by Helen G. Cruickshank. 191pp. 8¼ x 11. 23497-5 Pa. $6.00

AMERICAN WILDLIFE AND PLANTS, A. C. Martin, et al. Describes food habits of more than 1000 species of mammals, birds, fish. Special treatment of important food plants. Over 300 illustrations. 500pp. 5⅜ x 8½. 20793-5 Pa. $4.95

THE PEOPLE CALLED SHAKERS, Edward D. Andrews. Lifetime of research, definitive study of Shakers: origins, beliefs, practices, dances, social organization, furniture and crafts, impact on 19th-century USA, present heritage. Indispensable to student of American history, collector. 33 illustrations. 351pp. 5⅜ x 8½. 21081-2 Pa. $4.50

OLD NEW YORK IN EARLY PHOTOGRAPHS, Mary Black. New York City as it was in 1853-1901, through 196 wonderful photographs from N.-Y. Historical Society. Great Blizzard, Lincoln's funeral procession, great buildings. 228pp. 9 x 12. 22907-6 Pa. $8.95

MR. LINCOLN'S CAMERA MAN: MATHEW BRADY, Roy Meredith. Over 300 Brady photos reproduced directly from original negatives, photos. Jackson, Webster, Grant, Lee, Carnegie, Barnum; Lincoln; Battle Smoke, Death of Rebel Sniper, Atlanta Just After Capture. Lively commentary. 368pp. 8⅜ x 11¼. 23021-X Pa. $8.95

TRAVELS OF WILLIAM BARTRAM, William Bartram. From 1773-8, Bartram explored Northern Florida, Georgia, Carolinas, and reported on wild life, plants, Indians, early settlers. Basic account for period, entertaining reading. Edited by Mark Van Doren. 13 illustrations. 141pp. 5⅜ x 8½. 20013-2 Pa. $5.00

THE GENTLEMAN AND CABINET MAKER'S DIRECTOR, Thomas Chippendale. Full reprint, 1762 style book, most influential of all time; chairs, tables, sofas, mirrors, cabinets, etc. 200 plates, plus 24 photographs of surviving pieces. 249pp. 9⅞ x 12¾. 21601-2 Pa. $7.95

AMERICAN CARRIAGES, SLEIGHS, SULKIES AND CARTS, edited by Don H. Berkebile. 168 Victorian illustrations from catalogues, trade journals, fully captioned. Useful for artists. Author is Assoc. Curator, Div. of Transportation of Smithsonian Institution. 168pp. 8½ x 9½. 23328-6 Pa. $5.00

YUCATAN BEFORE AND AFTER THE CONQUEST, Diego de Landa. First English translation of basic book in Maya studies, the only significant account of Yucatan written in the early post-Conquest era. Translated by distinguished Maya scholar William Gates. Appendices, introduction, 4 maps and over 120 illustrations added by translator. 162pp. 5⅜ x 8½.
23622-6 Pa. $3.00

THE MALAY ARCHIPELAGO, Alfred R. Wallace. Spirited travel account by one of founders of modern biology. Touches on zoology, botany, ethnography, geography, and geology. 62 illustrations, maps. 515pp. 5⅜ x 8½.
20187-2 Pa. $6.95

THE DISCOVERY OF THE TOMB OF TUTANKHAMEN, Howard Carter, A. C. Mace. Accompany Carter in the thrill of discovery, as ruined passage suddenly reveals unique, untouched, fabulously rich tomb. Fascinating account, with 106 illustrations. New introduction by J. M. White. Total of 382pp. 5⅜ x 8½. (Available in U.S. only)   23500-9 Pa. $4.00

THE WORLD'S GREATEST SPEECHES, edited by Lewis Copeland and Lawrence W. Lamm. Vast collection of 278 speeches from Greeks up to present. Powerful and effective models; unique look at history. Revised to 1970. Indices. 842pp. 5⅜ x 8½.   20468-5 Pa. $8.95

THE 100 GREATEST ADVERTISEMENTS, Julian Watkins. The priceless ingredient; His master's voice; 99 44/100% pure; over 100 others. How they were written, their impact, etc. Remarkable record. 130 illustrations. 233pp. 7⅞ x 10 3/5.   20540-1 Pa. $5.95

CRUICKSHANK PRINTS FOR HAND COLORING, George Cruickshank. 18 illustrations, one side of a page, on fine-quality paper suitable for watercolors. Caricatures of people in society (c. 1820) full of trenchant wit. Very large format. 32pp. 11 x 16.   23684-6 Pa. $5.00

THIRTY-TWO COLOR POSTCARDS OF TWENTIETH-CENTURY AMERICAN ART, Whitney Museum of American Art. Reproduced in full color in postcard form are 31 art works and one shot of the museum. Calder, Hopper, Rauschenberg, others. Detachable. 16pp. 8¼ x 11.
23629-3 Pa. $3.00

MUSIC OF THE SPHERES: THE MATERIAL UNIVERSE FROM ATOM TO QUASAR SIMPLY EXPLAINED, Guy Murchie. Planets, stars, geology, atoms, radiation, relativity, quantum theory, light, antimatter, similar topics. 319 figures. 664pp. 5⅜ x 8½.
21809-0, 21810-4 Pa., Two-vol. set $11.00

EINSTEIN'S THEORY OF RELATIVITY, Max Born. Finest semi-technical account; covers Einstein, Lorentz, Minkowski, and others, with much detail, much explanation of ideas and math not readily available elsewhere on this level. For student, non-specialist. 376pp. 5⅜ x 8½.
60769-0 Pa. $4.50

THE EARLY WORK OF AUBREY BEARDSLEY, Aubrey Beardsley. 157 plates, 2 in color: *Manon Lescaut, Madame Bovary, Morte Darthur, Salome,* other. Introduction by H. Marillier. 182pp. 8⅛ x 11. 21816-3 Pa. $4.50

THE LATER WORK OF AUBREY BEARDSLEY, Aubrey Beardsley. Exotic masterpieces of full maturity: *Venus and Tannhauser, Lysistrata, Rape of the Lock, Volpone,* Savoy material, etc. 174 plates, 2 in color. 186pp. 8⅛ x 11.   21817-1 Pa. $5.95

THOMAS NAST'S CHRISTMAS DRAWINGS, Thomas Nast. Almost all Christmas drawings by creator of image of Santa Claus as we know it, and one of America's foremost illustrators and political cartoonists. 66 illustrations. 3 illustrations in color on covers. 96pp. 8⅜ x 11¼.
                                                            23660-9 Pa. $3.50

THE DORÉ ILLUSTRATIONS FOR DANTE'S DIVINE COMEDY, Gustave Doré. All 135 plates from Inferno, Purgatory, Paradise; fantastic tortures, infernal landscapes, celestial wonders. Each plate with appropriate (translated) verses. 141pp. 9 x 12.   23231-X Pa. $4.50

DORÉ'S ILLUSTRATIONS FOR RABELAIS, Gustave Doré. 252 striking illustrations of *Gargantua and Pantagruel* books by foremost 19th-century illustrator. Including 60 plates, 192 delightful smaller illustrations. 153pp. **9 x 12.**   23656-0 Pa. $5.00

LONDON: A PILGRIMAGE, Gustave Doré, Blanchard Jerrold. Squalor, riches, misery, beauty of mid-Victorian metropolis; 55 wonderful plates, 125 other illustrations, full social, cultural text by Jerrold. 191pp. of text. 9⅜ x 12¼.   22306-X Pa. $7.00

THE RIME OF THE ANCIENT MARINER, Gustave Doré, S. T. Coleridge. Dore's finest work, 34 plates capture moods, subtleties of poem. Full text. Introduction by Millicent Rose. 77pp. 9¼ x 12.   22305-1 Pa. $3.50

THE DORE BIBLE ILLUSTRATIONS, Gustave Doré. All wonderful, detailed plates: Adam and Eve, Flood, Babylon, Life of Jesus, etc. Brief King James text with each plate. Introduction by Millicent Rose. 241 plates. 241pp. 9 x 12.   23004-X Pa. $6.00

THE COMPLETE ENGRAVINGS, ETCHINGS AND DRYPOINTS OF ALBRECHT DURER. "Knight, Death and Devil"; "Melencolia," and more—all Dürer's known works in all three media, including 6 works formerly attributed to him. 120 plates. 235pp. 8⅜ x 11¼.
                                                            22851-7 Pa. $6.50

MECHANICK EXERCISES ON THE WHOLE ART OF PRINTING, Joseph Moxon. First complete book (1683-4) ever written about typography, a compendium of everything known about printing at the latter part of 17th century. Reprint of 2nd (1962) Oxford Univ. Press edition. 74 illustrations. Total of 550pp. 6⅛ x 9¼.   23617-X Pa. $7.95

THE COMPLETE WOODCUTS OF ALBRECHT DURER, edited by Dr. W. Kurth. 346 in all: "Old Testament," "St. Jerome," "Passion," "Life of Virgin," Apocalypse," many others. Introduction by Campbell Dodgson. 285pp. 8½ x 12¼. 21097-9 Pa. $7.50

DRAWINGS OF ALBRECHT DURER, edited by Heinrich Wolfflin. 81 plates show development from youth to full style. Many favorites; many new. Introduction by Alfred Werner. 96pp. 8⅛ x 11. 22352-3 Pa. $5.00

THE HUMAN FIGURE, Albrecht Dürer. Experiments in various techniques—stereometric, progressive proportional, and others. Also life studies that rank among finest ever done. Complete reprinting of *Dresden Sketchbook*. 170 plates. 355pp. 8⅜ x 11¼. 21042-1 Pa. $7.95

OF THE JUST SHAPING OF LETTERS, Albrecht Dürer. Renaissance artist explains design of Roman majuscules by geometry, also Gothic lower and capitals. Grolier Club edition. 43pp. 7⅞ x 10¾ 21306-4 Pa. $3.00

TEN BOOKS ON ARCHITECTURE, Vitruvius. The most important book ever written on architecture. Early Roman aesthetics, technology, classical orders, site selection, all other aspects. Stands behind everything since. Morgan translation. 331pp. 5⅜ x 8½. 20645-9 Pa. $4.50

THE FOUR BOOKS OF ARCHITECTURE, Andrea Palladio. 16th-century classic responsible for Palladian movement and style. Covers classical architectural remains, Renaissance revivals, classical orders, etc. 1738 Ware English edition. Introduction by A. Placzek. 216 plates. 110pp. of text. 9½ x 12¾. 21308-0 Pa. $10.00

HORIZONS, Norman Bel Geddes. Great industrialist stage designer, "father of streamlining," on application of aesthetics to transportation, amusement, architecture, etc. 1932 prophetic account; function, theory, specific projects. 222 illustrations. 312pp. 7⅞ x 10¾. 23514-9 Pa. $6.95

FRANK LLOYD WRIGHT'S FALLINGWATER, Donald Hoffmann. Full, illustrated story of conception and building of Wright's masterwork at Bear Run, Pa. 100 photographs of site, construction, and details of completed structure. 112pp. 9¼ x 10. 23671-4 Pa. $5.50

THE ELEMENTS OF DRAWING, John Ruskin. Timeless classic by great Viltorian; starts with basic ideas, works through more difficult. Many practical exercises. 48 illustrations. Introduction by Lawrence Campbell. 228pp. 5⅜ x 8½. 22730-8 Pa. $3.75

GIST OF ART, John Sloan. Greatest modern American teacher, Art Students League, offers innumerable hints, instructions, guided comments to help you in painting. Not a formal course. 46 illustrations. Introduction by Helen Sloan. 200pp. 5⅜ x 8½. 23435-5 Pa. $4.00

THE ANATOMY OF THE HORSE, George Stubbs. Often considered the great masterpiece of animal anatomy. Full reproduction of 1766 edition, plus prospectus; original text and modernized text. 36 plates. Introduction by Eleanor Garvey. 121pp. 11 x 14¾.   23402-9 Pa. $6.00

BRIDGMAN'S LIFE DRAWING, George B. Bridgman. More than 500 illustrative drawings and text teach you to abstract the body into its major masses, use light and shade, proportion; as well as specific areas of anatomy, of which Bridgman is master. 192pp. 6½ x 9¼. (Available in U.S. only)
22710-3 Pa. $3.50

ART NOUVEAU DESIGNS IN COLOR, Alphonse Mucha, Maurice Verneuil, Georges Auriol. Full-color reproduction of *Combinaisons ornementales* (c. 1900) by Art Nouveau masters. Floral, animal, geometric, interlacings, swashes—borders, frames, spots—all incredibly beautiful. 60 plates, hundreds of designs. 9⅜ x 8-1/16.   22885-1 Pa. $4.00

FULL-COLOR FLORAL DESIGNS IN THE ART NOUVEAU STYLE, E. A. Seguy. 166 motifs, on 40 plates, from *Les fleurs et leurs applications decoratives* (1902): borders, circular designs, repeats, allovers, "spots." All in authentic Art Nouveau colors. 48pp. 9⅜ x 12¼.
23439-8 Pa. $5.00

A DIDEROT PICTORIAL ENCYCLOPEDIA OF TRADES AND IN-DUSTRY, edited by Charles C. Gillispie. 485 most interesting plates from the great French Encyclopedia of the 18th century show hundreds of working figures, artifacts, process, land and cityscapes; glassmaking, paper-making, metal extraction, construction, weaving, making furniture, clothing, wigs, dozens of other activities. Plates fully explained. 920pp. 9 x 12.
22284-5, 22285-3 Clothbd., Two-vol. set $40.00

HANDBOOK OF EARLY ADVERTISING ART, Clarence P. Hornung. Largest collection of copyright-free early and antique advertising art ever compiled. Over 6,000 illustrations, from Franklin's time to the 1890's for special effects, novelty. Valuable source, almost inexhaustible.
Pictorial Volume. Agriculture, the zodiac, animals, autos, birds, Christmas, fire engines, flowers, trees, musical instruments, ships, games and sports, much more. Arranged by subject matter and use. 237 plates. 288pp. 9 x 12.
20122-8 Clothbd. $14.50

Typographical Volume. Roman and Gothic faces ranging from 10 point to 300 point, "Barnum," German and Old English faces, script, logotypes, scrolls and flourishes, 1115 ornamental initials, 67 complete alphabets, more. 310 plates. 320pp. 9 x 12.   20123-6 Clothbd. $15.00

CALLIGRAPHY (CALLIGRAPHIA LATINA), J. G. Schwandner. High point of 18th-century ornamental calligraphy. Very ornate initials, scrolls, borders, cherubs, birds, lettered examples. 172pp. 9 x 13.
20475-8 Pa. $7.00

ART FORMS IN NATURE, Ernst Haeckel. Multitude of strangely beautiful natural forms: Radiolaria, Foraminifera, jellyfishes, fungi, turtles, bats, etc. All 100 plates of the 19th-century evolutionist's *Kunstformen der Natur* (1904). 100pp. 9⅜ x 12¼.   22987-4 Pa. $5.00

CHILDREN: A PICTORIAL ARCHIVE FROM NINETEENTH-CENTURY SOURCES, edited by Carol Belanger Grafton. 242 rare, copyright-free wood engravings for artists and designers. Widest such selection available. All illustrations in line. 119pp. 8⅜ x 11¼.

23694-3 Pa. $4.00

WOMEN: A PICTORIAL ARCHIVE FROM NINETEENTH-CENTURY SOURCES, edited by Jim Harter. 391 copyright-free wood engravings for artists and designers selected from rare periodicals. Most extensive such collection available. All illustrations in line. 128pp. 9 x 12.

23703-6 Pa. $4.50

ARABIC ART IN COLOR, Prisse d'Avennes. From the greatest ornamentalists of all time—50 plates in color, rarely seen outside the Near East, rich in suggestion and stimulus. Includes 4 plates on covers. 46pp. 9⅜ x 12¼.   23658-7 Pa. $6.00

AUTHENTIC ALGERIAN CARPET DESIGNS AND MOTIFS, edited by June Beveridge. Algerian carpets are world famous. Dozens of geometrical motifs are charted on grids, color-coded, for weavers, needleworkers, craftsmen, designers. 53 illustrations plus 4 in color. 48pp. 8¼ x 11. (Available in U.S. only)   23650-1 Pa. $1.75

DICTIONARY OF AMERICAN PORTRAITS, edited by Hayward and Blanche Cirker. 4000 important Americans, earliest times to 1905, mostly in clear line. Politicians, writers, soldiers, scientists, inventors, industrialists, Indians, Blacks, women, outlaws, etc. Identificatory information. 756pp. 9¼ x 12¾.   21823-6 Clothbd. $40.00

HOW THE OTHER HALF LIVES, Jacob A. Riis. Journalistic record of filth, degradation, upward drive in New York immigrant slums, shops, around 1900. New edition includes 100 original Riis photos, monuments of early photography. 233pp. 10 x 7⅞.   22012-5 Pa. $7.00

NEW YORK IN THE THIRTIES, Berenice Abbott. Noted photographer's fascinating study of city shows new buildings that have become famous and old sights that have disappeared forever. Insightful commentary. 97 photographs. 97pp. 11⅜ x 10.   22967-X Pa. $5.00

MEN AT WORK, Lewis W. Hine. Famous photographic studies of construction workers, railroad men, factory workers and coal miners. New supplement of 18 photos on Empire State building construction. New introduction by Jonathan L. Doherty. Total of 69 photos. 63pp. 8 x 10¾.

23475-4 Pa. $3.00

THE DEPRESSION YEARS AS PHOTOGRAPHED BY ARTHUR ROTH-STEIN, Arthur Rothstein. First collection devoted entirely to the work of outstanding 1930s photographer: famous dust storm photo, ragged children, unemployed, etc. 120 photographs. Captions. 119pp. 9¼ x 10¾.
23590-4 Pa. $5.00

CAMERA WORK: A PICTORIAL GUIDE, Alfred Stieglitz. All 559 illustrations and plates from the most important periodical in the history of art photography, Camera Work (1903-17). Presented four to a page, reduced in size but still clear, in strict chronological order, with complete captions. Three indexes. Glossary. Bibliography. 176pp. 8⅜ x 11¼.
23591-2 Pa. $6.95

ALVIN LANGDON COBURN, PHOTOGRAPHER, Alvin L. Coburn. Revealing autobiography by one of greatest photographers of 20th century gives insider's version of Photo-Secession, plus comments on his own work. 77 photographs by Coburn. Edited by Helmut and Alison Gernsheim. 160pp. 8⅛ x 11.
23685-4 Pa. $6.00

NEW YORK IN THE FORTIES, Andreas Feininger. 162 brilliant photographs by the well-known photographer, formerly with Life magazine, show commuters, shoppers, Times Square at night, Harlem nightclub, Lower East Side, etc. Introduction and full captions by John von Hartz. 181pp. 9¼ x 10¾.
23585-8 Pa. $6.00

GREAT NEWS PHOTOS AND THE STORIES BEHIND THEM, John Faber. Dramatic volume of 140 great news photos, 1855 through 1976, and revealing stories behind them, with both historical and technical information. Hindenburg disaster, shooting of Oswald, nomination of Jimmy Carter, etc. 160pp. 8¼ x 11.
23667-6 Pa. $5.00

THE ART OF THE CINEMATOGRAPHER, Leonard Maltin. Survey of American cinematography history and anecdotal interviews with 5 masters—Arthur Miller, Hal Mohr, Hal Rosson, Lucien Ballard, and Conrad Hall. Very large selection of behind-the-scenes production photos. 105 photographs. Filmographies. Index. Originally Behind the Camera. 144pp. 8¼ x 11.
23686-2 Pa. $5.00

DESIGNS FOR THE THREE-CORNERED HAT (LE TRICORNE), Pablo Picasso. 32 fabulously rare drawings—including 31 color illustrations of costumes and accessories—for 1919 production of famous ballet. Edited by Parmenia Migel, who has written new introduction. 48pp. 9⅜ x 12¼. (Available in U.S. only)
23709-5 Pa. $5.00

NOTES OF A FILM DIRECTOR, Sergei Eisenstein. Greatest Russian filmmaker explains montage, making of Alexander Nevsky, aesthetics; comments on self, associates, great rivals (Chaplin), similar material. 78 illustrations. 240pp. 5⅜ x 8½.
22392-2 Pa. $4.50

HOLLYWOOD GLAMOUR PORTRAITS, edited by John Kobal. 145 photos capture the stars from 1926-49, the high point in portrait photography. Gable, Harlow, Bogart, Bacall, Hedy Lamarr, Marlene Dietrich, Robert Montgomery, Marlon Brando, Veronica Lake; 94 stars in all. Full background on photographers, technical aspects, much more. Total of 160pp. 8⅜ x 11¼.   23352-9 Pa. $6.00

THE NEW YORK STAGE: FAMOUS PRODUCTIONS IN PHOTO-GRAPHS, edited by Stanley Appelbaum. 148 photographs from Museum of City of New York show 142 plays, 1883-1939. *Peter Pan, The Front Page, Dead End, Our Town,* O'Neill, hundreds of actors and actresses, etc. Full indexes. 154pp. 9½ x 10.   23241-7 Pa. $6.00

MASTERS OF THE DRAMA, John Gassner. Most comprehensive history of the drama, every tradition from Greeks to modern Europe and America, including Orient. Covers 800 dramatists, 2000 plays; biography, plot summaries, criticism, theatre history, etc. 77 illustrations. 890pp. 5⅜ x 8½.
20100-7 Clothbd. $10.00

THE GREAT OPERA STARS IN HISTORIC PHOTOGRAPHS, edited by James Camner. 343 portraits from the 1850s to the 1940s: Tamburini, Mario, Caliapin, Jeritza, Melchior, Melba, Patti, Pinza, Schipa, Caruso, Farrar, Steber, Gobbi, and many more—270 performers in all. Index. 199pp. 8⅜ x 11¼.   23575-0 Pa. $6.50

J. S. BACH, Albert Schweitzer. Great full-length study of Bach, life, background to music, music, by foremost modern scholar. Ernest Newman translation. 650 musical examples. Total of 928pp. 5⅜ x 8½. (Available in U.S. only)   21631-4, 21632-2 Pa., Two-vol. set $10.00

COMPLETE PIANO SONATAS, Ludwig van Beethoven. All sonatas in the fine Schenker edition, with fingering, analytical material. One of best modern editions. Total of 615pp. 9 x 12. (Available in U.S. only)
23134-8, 23135-6 Pa., Two-vol. set $15.00

KEYBOARD MUSIC, J. S. Bach. Bach-Gesellschaft edition. For harpsichord, piano, other keyboard instruments. English Suites, French Suites, Six Partitas, Goldberg Variations, Two-Part Inventions, Three-Part Sinfonias. 312pp. 8⅛ x 11. (Available in U.S. only)   22360-4 Pa. $6.95

FOUR SYMPHONIES IN FULL SCORE, Franz Schubert. Schubert's four most popular symphonies: No. 4 in C Minor ("Tragic"); No. 5 in B-flat Major; No. 8 in B Minor ("Unfinished"); No. 9 in C Major ("Great"). Breitkopf & Hartel edition. Study score. 261pp. 9⅜ x 12¼.
23681-1 Pa. $6.50

THE AUTHENTIC GILBERT & SULLIVAN SONGBOOK, W. S. Gilbert, A. S. Sullivan. Largest selection available; 92 songs, uncut, original keys, in piano rendering approved by Sullivan. Favorites and lesser-known fine numbers. Edited with plot synopses by James Spero. 3 illustrations. 399pp. 9 x 12.   23482-7 Pa. $7.95

AMERICAN ANTIQUE FURNITURE, Edgar G. Miller, Jr. The basic coverage of all American furniture before 1840: chapters per item chronologically cover all types of furniture, with more than 2100 photos. Total of 1106pp. 7⅞ x 10¾.   21599-7, 21600-4 Pa., Two-vol. set $17.90

ILLUSTRATED GUIDE TO SHAKER FURNITURE, Robert Meader. Director, Shaker Museum, Old Chatham, presents up-to-date coverage of all furniture and appurtenances, with much on local styles not available elsewhere. 235 photos. 146pp. 9 x 12.   22819-3 Pa. $6.00

ORIENTAL RUGS, ANTIQUE AND MODERN, Walter A. Hawley. Persia, Turkey, Caucasus, Central Asia, China, other traditions. Best general survey of all aspects: styles and periods, manufacture, uses, symbols and their interpretation, and identification. 96 illustrations, 11 in color. 320pp. 6⅛ x 9¼.   22366-3 Pa. $6.95

CHINESE POTTERY AND PORCELAIN, R. L. Hobson. Detailed descriptions and analyses by former Keeper of the Department of Oriental Antiquities and Ethnography at the British Museum. Covers hundreds of pieces from primitive times to 1915. Still the standard text for most periods. 136 plates, 40 in full color. Total of 750pp. 5⅜ x 8½.
23253-0 Pa. $10.00

THE WARES OF THE MING DYNASTY, R. L. Hobson. Foremost scholar examines and illustrates many varieties of Ming (1368-1644). Famous blue and white, polychrome, lesser-known styles and shapes. 117 illustrations, 9 full color, of outstanding pieces. Total of 263pp. 6⅛ x 9¼. (Available in U.S. only)   23652-8 Pa. $6.00

Prices subject to change without notice.

Available at your book dealer or write for free catalogue to Dept. GI, Dover Publications, Inc., 180 Varick St., N.Y., N.Y. 10014. Dover publishes more than 175 books each year on science, elementary and advanced mathematics, biology, music, art, literary history, social sciences and other areas.